Communications Satellites

Communications Satellites

The technology of space communications

LARRY BLONSTEIN
BSc(Eng), MIMechE, MIEE, MRAeS

A HALSTED PRESS BOOK

JOHN WILEY & SONS, INC.
New York

Published in the USA by
Halsted Press, a division of
John Wiley & Sons, Inc., New York

ISBN 0470-20825-2

Printed in Great Britain

This book is dedicated to all the engineers throughout the world who have made space communications an economic reality

Publisher's note

This book was already with the typesetter when we received news of Larry Blonstein's sudden death. All who have contributed to its preparation are very sorry that he could not see it in its final form and share our pride in its publication.

The Blonstein Prize in Mechanical Engineering

As a memorial to Larry Blonstein, and in recognition of his achievements, his ex-colleagues and friends are contributing to a prize that is to be instituted at University College London, where he took his degree in mechanical engineering.

Preface

This book is for those who know little or nothing about communications satellites but who would like to know more, for a variety of reasons. There are those who are entering the world of space communications for the first time and who need to understand the fundamental aspects of communicating via space so that they can apply the techniques to their commercial or industrial requirements. There are others, who are already in the business of communications, who need to compare the advantages and disadvantages of distributing their information by space or by terrestrial means. There are those who already use space communications but who need to understand the jargon used by the world's manufacturers of spacecraft and Earth stations; and there are those who are just interested in the fascinating range of technologies that are applied in satellites, which are changing the face of communications across the world.

No previous knowledge of the technologies and mathematics involved is assumed. The book starts with simple technical analogies and proceeds, step by step, to a point where the reader can calculate, for himself, the essential technical characteristics of a space-linked communications system. Every step is explained on the way, including the most basic mathematical processes that are required, the reasoning behind the various techniques that need to be selected, and the ways of quantifying the communications parameters that determine the quality of a voice, data or video service. It also presents information on the economics of space communications, on the widening practice of digital encryption for commercial security, and on satellite systems that are currently in operation or that are due to come into service soon.

This is not a textbook. It cannot, within its size limitations, penetrate the depths of the vast range of technologies that are applied in communications satellites. It is a primer, which it is hoped may lead the interested reader to those texts that do deal fully with all the technologies of space and communications. As a primer, it happily and unashamedly mixes its units; to the English or American ear, 'miles per hour' is still more meaningful than 'kilometres per hour', but in matters of communications engineering it is appropriate to follow international practice and use metric units.

J. L. B.
October 1986

Contents

List of photographs

Acknowledgements

All the material in this book has been gathered together from my own experiences in the worlds of satellites and space communications and from those of my many friends and colleagues in the space industries and authorities in the USA and throughout Europe. They are far too numerous to name, but I must thank particularly Dr Richard Barnett who – unlike me, a Jack-of-all-trades in the arena of space – is an expert communications engineer, and who kept me on the right track in my meanderings in the fields of communications.

I thank, too, Martin Hellman of Stanford University, whose work I drew on to produce the chapter on encryption, an aspect of communications that holds a delightful fascination for anyone with a mathematical mind.

And again I thank my wife, Lys de Bray, also an author, who understands writers' needs for sustenance at strange times of the night.

Introduction

When a television studio engineer in Rome picks up a telephone to order a satellite link for a one-hour TV programme to London and New York, he does so without a second thought. To him, this is routine, and he is not even aware that his telephone call goes by satellite too. It is equally routine for the banker in Washington who wants to send financial data to every branch across the USA, instantaneously; for the shipping agent who wants to call a vessel in mid-Atlantic; for the cable TV operator in a big city, who wants to add another channel to his programme catalogue; for the designer who wants to see, on his desk-top colour TV, the piece of equipment that a supplier is offering from a factory that may be 5, or 50, or 500, or 5000 miles away.

Every day sees an increase in the numbers of men and women, in commerce, industry and entertainment, who are using satellite links to conduct their business. Every day sees an increase in the numbers of Earth stations, with their white faces peering at different spots in the sky, on the roofs and in the car parks of factories, office blocks, TV studios and hotels. Soon, the gardens and house rooftops of Europe, too, will be studded with little antennas, less than 1 metre in diameter – already nearly two million home TV antennas are installed in the USA.

All this is because of communications satellites: because they have opened up new communications routes that cover vast areas instantaneously and that provide facilities that have never been available by terrestrial means. Satellites offer the ability to send voice, data, facsimile, videotext and colour TV to and from hundreds or even thousands of sites in a big corporate network; the ability to send TV programmes right across continents, and round the world for that matter, without having to book space on perhaps a dozen microwave or cable systems in a dozen countries; the ability of a corporation to be in entire control of its communications network – to change it as the business changes, immediately; the ability to exchange data between computers, at phenomenal speeds; to meet without travelling; to coalesce far-flung outposts into a coherent whole.

Most important of all, satellites offer the ability to save money. If communications satellites did not provide facilities that were cheaper than those equivalents that can be achieved by terrestrial circuits, no one would ever buy them. But buy them they do. Today, there are just over 100 communications satellites in orbit. Of these, nearly eighty are dedicated to users in the worlds of commerce, industry and entertainment. These commercial communications satellites are earning profits for their owners, and saving the communications costs of their users at the same time. Reduction in communications costs means more profit for the users too, which is why the numbers of commercial communications satellites and Earth stations continue to grow.

This book is about those satellites. It is about how they work and how they are used. It is about Earth stations as well, because they are an essential part of the end-to-end links between users. For anyone whose job it is, or might be, to take responsibility for leading his or her business into the world of space communications, the book takes a step-by-step approach to the process of selecting the right satellite and Earth stations for a range of applications. It explains not only the technologies and the mathematics involved, but also the meanings of the many jargon terms that have crept into the language of space communications. Its aim is to help today's users, and the users of tomorrow, to understand the fundamental aspects of space communications so that, when they meet the experts of the world's satellite and Earth station industries, *they know what they are talking about.*

1

Getting a crate into orbit

If you take a crate to the top of a cliff and just tip it over the edge, it will fall vertically and crash on the rocks below (Figure 1).

This statement is so self-evident that it appears trivial, but it is fundamental to the understanding of satellite orbits. The falling crate is demonstrating one of the physical laws of the Universe – that all bodies are attracted to all other bodies under precise mathematical conditions. In this case, it is the Earth and the crate that are attracting each other, but as the mass of the Earth is billions of times more than that of the crate, it is the crate that goes rushing towards the Earth. As far as the crate 'knows', all the mass of the Earth is concentrated at its centre of gravity, 4000 miles below the surface. That is where the crate is trying to go when it hits the rocks.

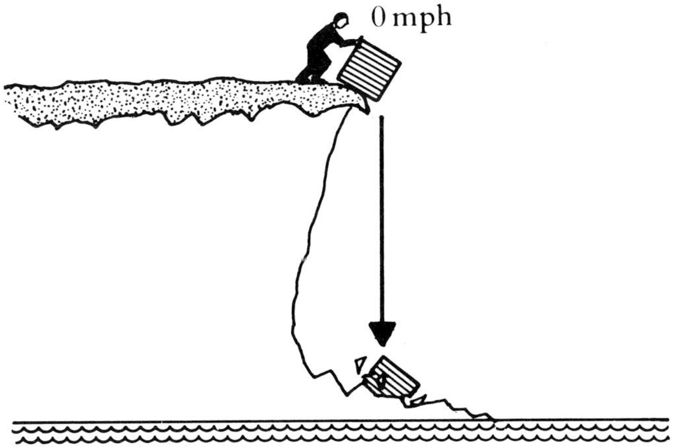

Figure 1 *A crate tipped off a cliff falls vertically to the centre of the Earth*

Suppose that you give the crate a push outwards from the cliff top: now it has a little horizontal speed. If you and your crate were alone out in

Figure 2 *Given a small forward speed, the crate flies out a little*

space together, your push would send the crate away from you in a straight line, and it would go on moving away from you in a straight line for ever. But you are not alone. Once again, the crate is attracted to the Earth's centre of gravity, and down it goes, but not vertically. It describes a parabolic path (Figure 2) that results from the speed that you gave it, which stays constant, and from the speed of falling, which increases continuously as the crate accelerates downwards under the constant pull of the Earth. Just before the crate hits the ground, it is falling almost vertically.

If you can find someone mad enough to drive a van at your crate so that it flies off the cliff top at, say, 50 mph, it will obviously travel farther (Figure 3). The crate is still trying to fall towards the Earth's centre of gravity, and if the Earth's pull has given it a downwards speed also of 50 mph at the end of its flight, it will fall into the sea at an angle of 45°.

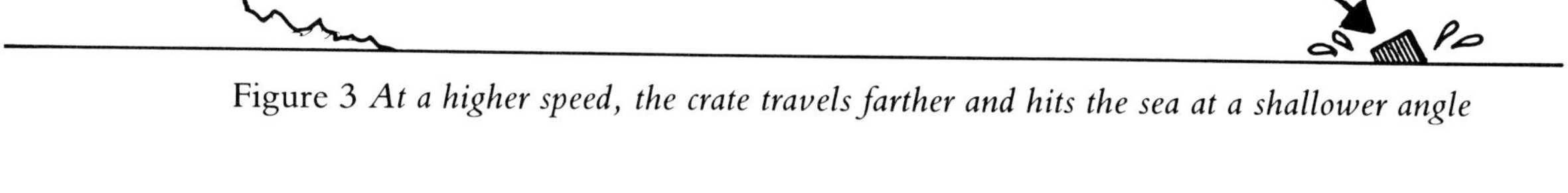

Figure 3 *At a higher speed, the crate travels farther and hits the sea at a shallower angle*

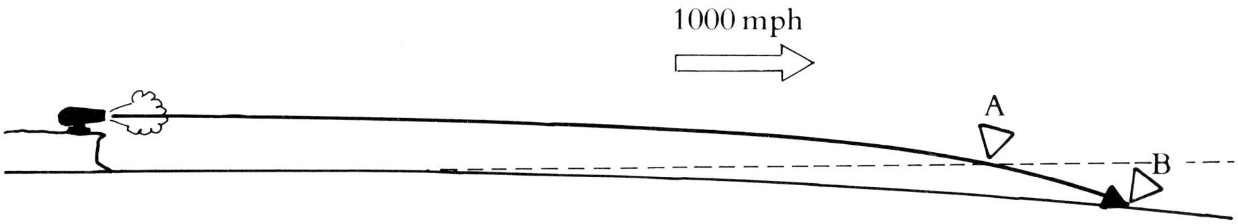

Figure 4 *At much higher speed, a new phenomenon appears. If the Earth were flat, the crate would hit the surface at point* A*. But the Earth's surface is curved, so it goes farther, to point* B

Clearly, the faster the crate flies off the cliff top, the farther it will go; but as the speed increases, a new phenomenon appears. Try firing the crate out of a circus cannon, so that it takes off at 1000 mph. It flies a long way out to sea, but farther than you would have expected, *because the Earth's surface is curved* (Figure 4). If the Earth were flat, the crate would fall into the sea at point A, but because the surface of the sea is curving away under the crate's trajectory, it hits the sea farther out, at point B.

Now we can begin to see how to get the crate into orbit. Try a still higher speed – 10 000 mph – by strapping the crate to a rocket (Figure 5).

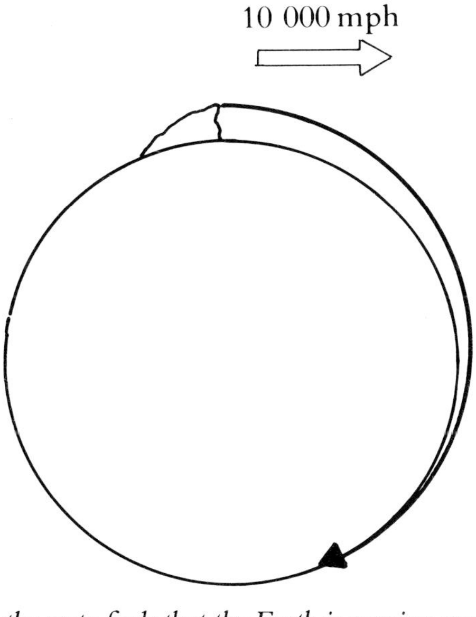

Figure 5 *At 10 000 mph, the crate finds that the Earth is curving away almost as quickly as it is falling towards it*

Now, the horizontal speed is so high that it almost overwhelms the downward speed caused by the Earth's pull, but not quite. The crate is still falling, but the Earth's surface is curving away under it as it falls. Eventually, it falls into the sea, almost horizontally, halfway round the world.

There must be a speed to get it to go all the way round. There is – 17 000 mph. This is the horizontal speed at which the crate still tries to fall towards the centre of the Earth but finds that the Earth's surface is curving away under it at the very rate at which it is trying to fall (Figure 6). The resulting path is a circle, and if it really were possible to fire crates off cliffs like this, you would have your crate hitting you in the back of the neck one-and-a-half hours later while you were standing on the cliff top wondering where it had gone!

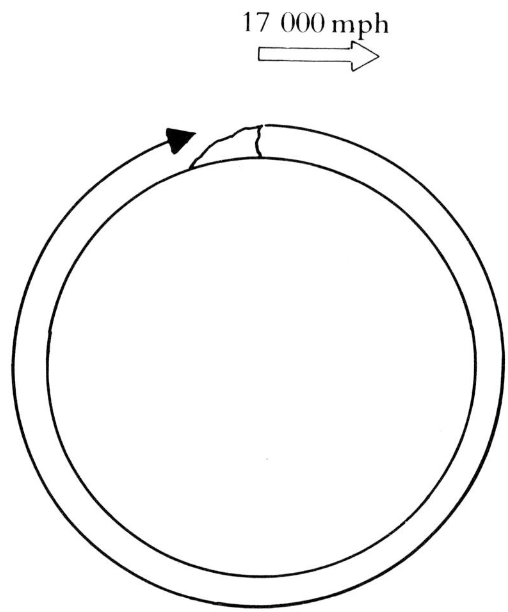

Figure 6 *At 17 000 mph, the falling crate never reaches the surface. It is in orbit*

In theory, you have got your crate into orbit, but in practice this is not possible at all. Air is the problem: you cannot fly objects through air at 17 000 mph. At sea level, air molecules cannot get out of the way of objects travelling faster than the speed of sound, which is about 720 mph. At higher speeds, they pack together to form dense waves – supersonic shock waves – that form up in front of the object and give rise to the air pressure increases that are heard as sonic 'booms'. It is not the booms that cause the problem – it is the heating of the molecules caused by the shock wave compression. This heat is transferred to the object's skin which, at

17 000 mph, will heat up to several thousand degrees and melt. This is why the underbelly and leading edges of the US Space Shuttle are covered with refractory tiles that can withstand such heat when it comes back into the Earth's atmosphere from space.

So, our crate cannot be flown round the world at 17 000 mph at cliff-top height. It must be taken up to an altitude where there is no air, or at least so little air that the few air molecules that get in the way do no harm to the crate. An altitude of 200 miles is adequate. At this height, the density of air molecules is so low that objects can safely be flown through them at 17 000 mph.

To get there, you put your crate on a vertical take-off rocket, which climbs on a carefully planned flight path until it is travelling horizontally – that is, parallel to the Earth's surface below it – at 17 000 mph (Figure 7).

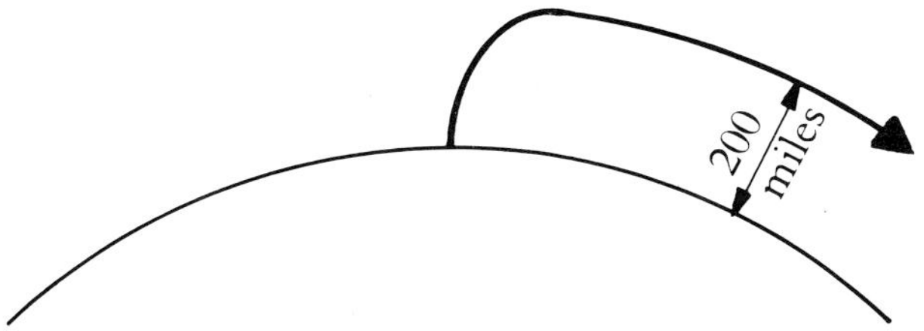

Figure 7 *To avoid air drag, satellites are lifted vertically by rockets to an altitude of about 200 miles before being sent on their circular paths*

At that point, the rocket nose, which has protected the crate on the climb through the atmosphere, opens, and the crate is ejected. It is in orbit, and it will stay in orbit for several years, falling around the Earth in its circular path. Eventually the slight drag of the residual atmosphere at that height causes the orbit to decay until the crate descends back into the atmosphere to burn up.

It is this 'falling round the Earth' that gives rise to the apparent 'weightlessness' of space. It is no surprise today to see on TV astronauts floating around inside the cabin of the Space Shuttle, taking 'space walks' outside, or even handling with obvious ease satellites that would weigh a couple of tons down on Earth. All this is simply because the Shuttle, the astronauts, the satellites, and all the other equipment on board, are falling round the Earth together. It is exactly as if you were in an elevator that was falling towards the ground because its lifting cable had broken. If you were carrying a suitcase, you could let it go and it would stay by your side, falling with you. It would appear to be weightless. If you were standing on a weighing machine in the elevator, it would also be falling with you, and its reading would show zero. You would appear to be weightless. There would appear to be no gravity. This is why the term *zero gravity* is often used in spaceflight. Of course, the Earth's gravitational pull is still there: it appears not to exist as long as you and everything around you go on falling together.

Zero gravity is exploited in the design of spacecraft – it permits devices that have to rotate to be mounted on very small, light bearings that would collapse under the weight of the devices on Earth. It permits fluids such as heat-transfer refrigerants to be moved in conditions that would be impossible on Earth, as described in Chapter 5.

At an altitude in the region of 150 to 400 miles, the crate will be in the company of hundreds of other satellites, travelling in all directions (Figure 8). Thousands of low-orbit satellites have been launched since the Russians put up the first one – *Sputnik* – in 1957, some of them lasting for only a few weeks, some for months and many for years. They are used for many purposes – scientific probing of the ionosphere, astronomy, land surveying, navigation, rescue, monitoring of crops and natural resources, and a variety of military applications that range from electronic surveillance and early warning to nuclear blast detection.

Low-orbit satellites are not used for communications, and Figure 8 shows why. To anyone on Earth, travelling on its surface from west to east as the Earth rotates once per day, all the satellites shown in Figure 8 will appear to 'rise' from one horizon and 'set' below another horizon about 30 minutes later. All of them, in that altitude region of 150 to 400 miles, are travelling at about 17 000 mph – their speeds do not differ much over this small range of heights – so all of them will take about 1½ hours to traverse the 24 000-mile circumference of the Earth. To someone on Earth, then, at point A, looking at a satellite that is on orbit B, the satellite will appear above the horizon in the north-east, climb to a maximum elevation about 15 minutes later, and then fall towards the south-west horizon, where it will 'set' after another 15 minutes. Next time round, 1½ hours later, the person on Earth will have moved 22° eastwards, and the satellite will 'rise' from a different location in the north-east.

Even though all these relative motions are predictable, and are accepted by low-orbit satellite users who have to track their spacecraft across the sky on each pass, they are not acceptable to those involved in communications. All communicators, be they telephone authorities, television companies or private businesses, want to be in contact 24 hours per day, not for just 30 minutes every 1½ hours. Further, they cannot be

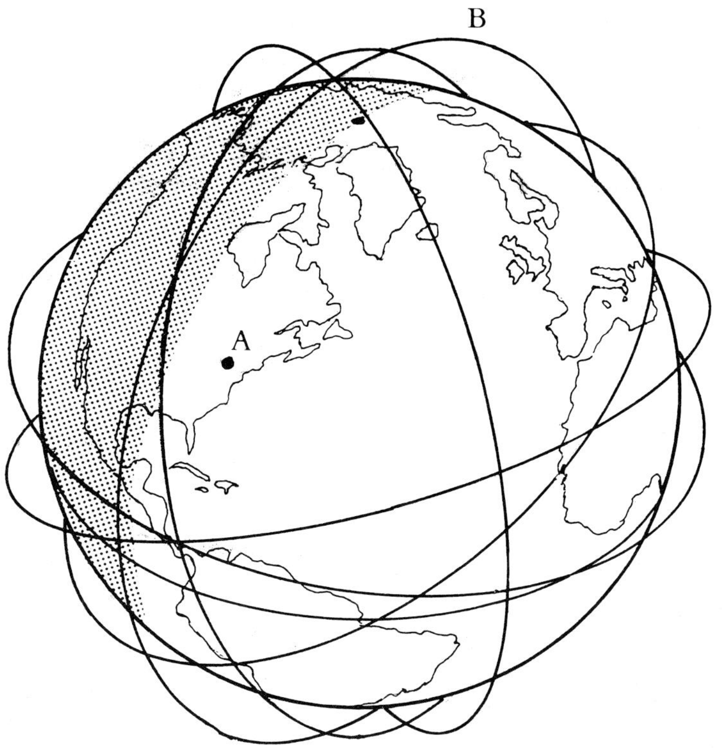

Figure 8 *Of the hundreds of satellites circling the Earth at low altitude, none is used for communications*

expected to invest in computer-driven tracking stations at every studio and factory and office. Such stations cost millions, and need teams of men on shiftwork around the clock to control them and to keep them maintained.

Our crate then, in its low $1\frac{1}{2}$-hour orbit, is not going to be much use there if it is to be used as a communications satellite. It has to be taken to another orbit, which cuts out the need for expensive tracking stations – the geosynchronous orbit.

2

The geosynchronous orbit

When a satellite is falling round the Earth on its circular path, its orbital speed is directly related to its altitude – in fact to its distance from the Earth's centre of gravity. The relationship is the result of two forces:

1 the Earth's gravitational attraction on the satellite, which tries to pull it inward towards the Earth's centre of gravity;
2 the opposing centrifugal force on the satellite, which tries to pull it outward.

Force 1 is proportional to the mass of the Earth, the mass of the satellite and the universal gravitational constant g, and inversely proportional to the square of the distance between the satellite and the Earth; i.e., if the distance is doubled, the force decreases by a factor of four. Force 2 is proportional to the mass of the satellite and the square of its orbital speed, and inversely proportional to its distance from the Earth.

All this reduces to a simple relationship between orbital speed and altitude, and thus to the time it will take for a satellite travelling at a particular speed to traverse its circular path at the matching altitude. The resulting relationship between time, speed and altitude is shown in Figure 9. Here we can see the theoretical time of $1\frac{1}{2}$ hours for a satellite travelling near the Earth's surface at 17 000 mph. As the satellite is taken up to higher altitudes, the gravitational attraction of the Earth is reduced, and the centrifugal balancing force can be lessened by reducing its speed. A rocket taking a satellite to 10 000 miles altitude, for instance, will need to send it on its way at only 9000 mph. At that speed, the satellite will take 9 hours to traverse its circular path at that height.

It was an Englishman, Arthur C. Clarke, who recognised a particular implication of this time–speed–altitude relationship. In 1945, he published an article drawing attention to the fact that, as Figure 9 shows, at an

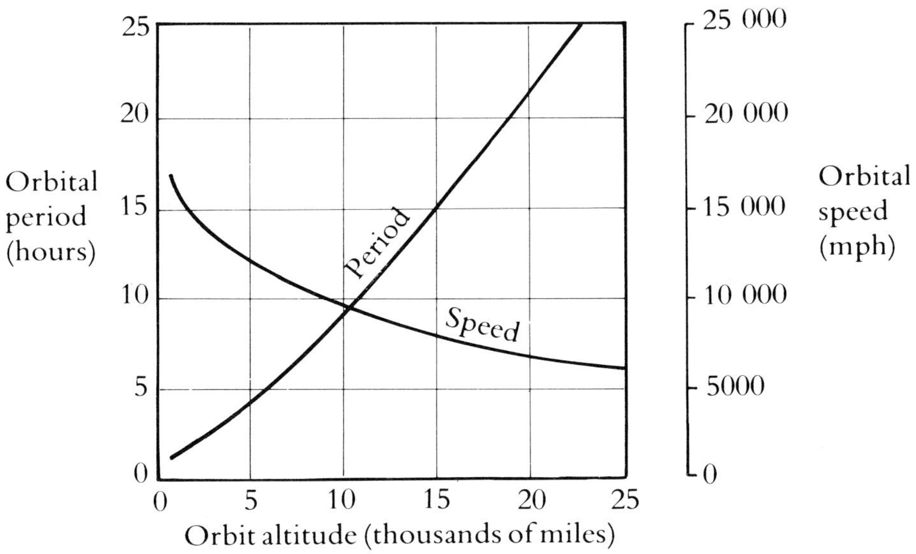

Figure 9 *Relationship between orbital period, speed and altitude*

altitude of 22 000 miles a satellite would take 24 hours to circle the Earth. Thus, he argued, because the Earth is also rotating once every 24 hours, a satellite at that height would appear to an observer on Earth to be stationary. That observer would not need to track the satellite: he could simply point his dish aerial at the satellite and lock it in place. Clarke went on to suggest that communications around the entire world could then be accomplished by placing three satellites out at 22 000 miles, equally spaced by 120°, on an orbital plane in line with the Equator.

Why in line with the Equator? Because any other orbital plane would not offer 'stationary' satellites. Figure 10(a) shows a 24-hour orbit, but with the satellite passing over the North and South Poles. Clearly, someone on Earth, travelling from west to east with the Earth's rotation, would see the satellite appearing from the north, travelling overhead and then disappearing again to the south. A full tracking station would be needed and the satellite would be in view for only a part of the day. Similarly, the inclined orbit shown in Figure 10(b), even though a 24-hour orbit, gives rise to the same problems. The orbit shown in Figure 11, while being potentially valuable, is not possible. It would be very convenient to have the orbit placed in line with 51° latitude, for instance, so that a satellite could be stationed permanently above London. However, such a satellite would not be falling around the centre of the

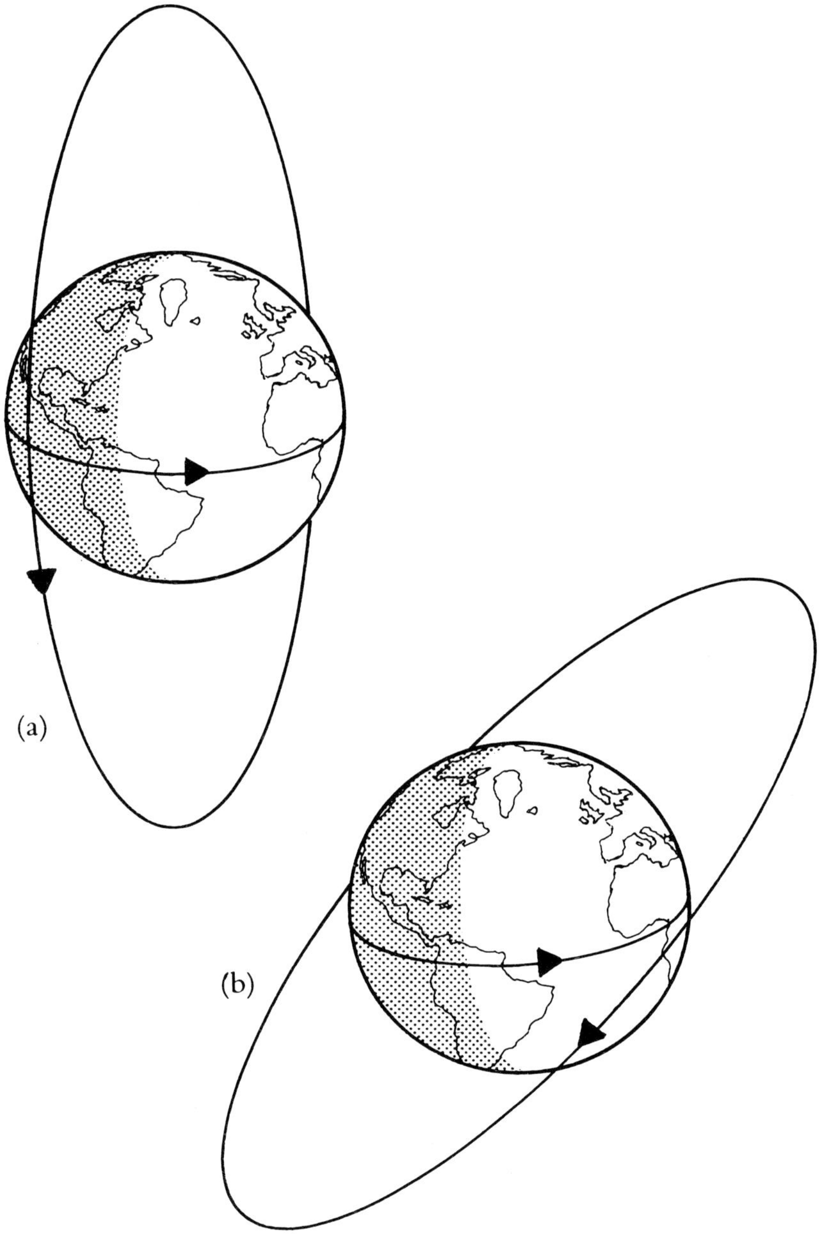

Figure 10(a) *A 24-hour orbit over the poles is no use for communications. The satellite and the user are travelling in different directions* (b) *An inclined orbit is no use either, for the same reason*

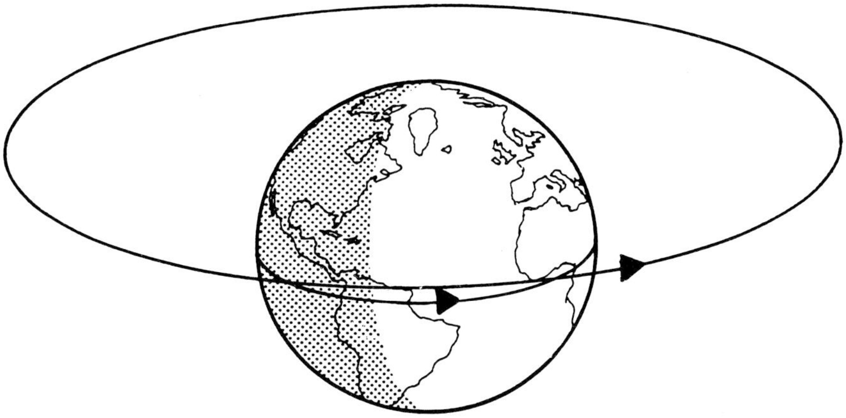

Figure 11 *This orbit is impossible – the satellite is not 'falling' towards the centre of the Earth*

Earth; it would be falling around a point some 3000 miles north of the Earth's centre of gravity, which is not possible.

The only place for the 24-hour orbit is in the plane of the Equator (Figure 12). Here, the satellite is travelling from east to west exactly in synchronism with people on the Earth below – thus the name *geosynchronous orbit.*

Clarke's identification of the geosynchronous orbit – the answer to economical communications via satellite – went unrecognised for years. It was not until 1965 – twenty years later – that the first geosynchronous communications satellite appeared. Called *Early Bird,* it was built by the Hughes Aircraft Company and launched from the USA. It carried 240 telephone circuits across the Atlantic. Today, geosynchronous satellites that can handle 30 000 circuits are becoming commonplace, and they are

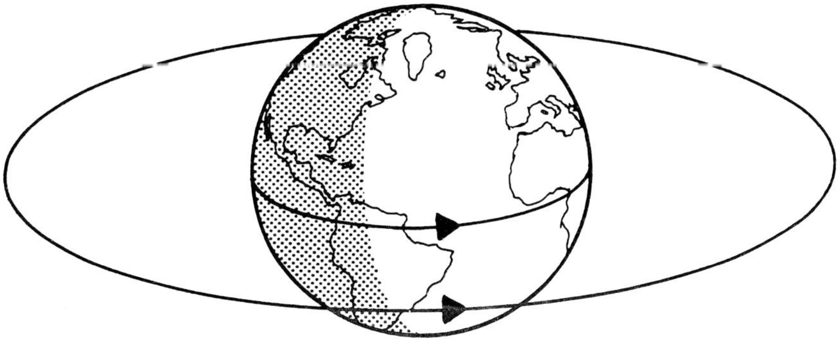

Figure 12 *Only this orbit, the geosynchronous orbit, is suitable for communications*

increasing in number (see Chapter 15) to the extent that countries are jostling for 'parking slots' in the geosynchronous arc where they can station their national satellites.

Today's communications traffic via geosynchronous satellites is a celestial monument to that revelation, 40 years ago, by Arthur Clarke, who is still happily producing thought-provoking concepts, books and films from his home in Sri Lanka.

3

From low orbit to geosynchronous altitude

Today, two different launching systems are used to put satellites into geosynchronous orbit – the Ariane rocket, built in France with subsystems supplied from several other European countries, and the Space Shuttle, built in the USA. Both launchers take the satellites to low orbit and then employ different means to put them into what is known as an elliptical transfer orbit.

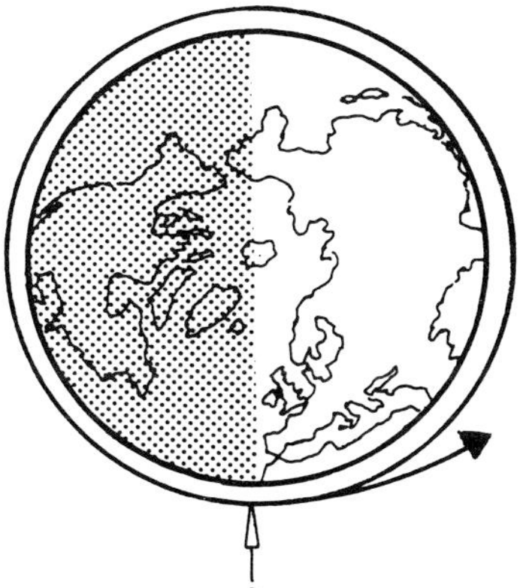

Figure 13 *An increase of speed at point* A *will throw the satellite out of its circular path*

An elliptical orbit is produced when the speed of the satellite (or our crate) is increased beyond that required to maintain a circular orbit. As we have seen, at about 200 miles altitude, a speed of 17 000 mph will keep the crate travelling in a circular path around the Earth (Figure 6). If the speed is increased to, say, 20 000 mph when the crate is passing through point A (Figure 13) it can be seen that this will unbalance the circular balancing equation and 'throw' the crate outward, away from the Earth. But, as

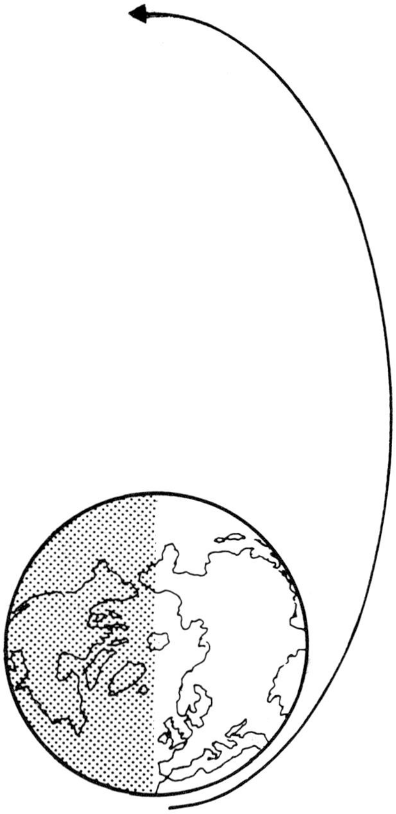

Figure 14 *If the speed at point* A *is increased from 17 000 mph to 22 800 mph, the satellite will climb on an elliptical path until it reaches an altitude of 22 000 miles*

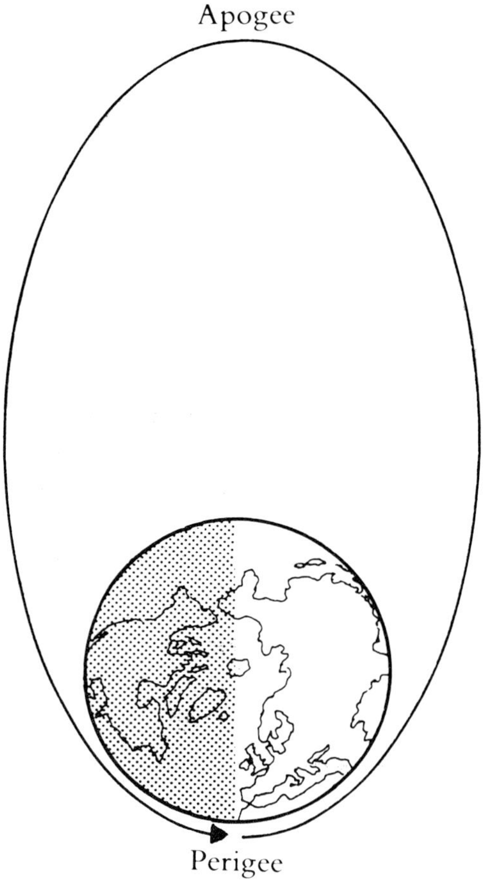

Figure 15 *If no other speed change is made, the satellite will continue to trace an elliptical path between apogee and perigee*

before, it is still being attracted back to the Earth's centre. The result of these two motions is an elliptical path, as seen in Figure 14. Eventually, the crate will climb to the high point of its path (the *apogee*) before falling back towards the Earth again. Because it still has forward speed at apogee, it will not fall vertically back to Earth, but will follow another elliptical path which is the exact mirror-image of its ascent, accelerating as it falls until its speed regains the original 20 000 mph at its lowest altitude (the *perigee* – Figure 15). The crate will go on repeating this elliptical orbit until another speed alteration is made to change the orbit shape.

If the speed at point A is increased to 22 800 mph, the apogee of the elliptical orbit becomes 22 000 miles, which is the height required for geosynchronous operations, but the forward speed of the crate is only 3000 mph, not the 7000 mph required to maintain a circular orbit. Thus the crate will stay in its elliptical transfer orbit, with an apogee of 22 000 miles and a perigee of 200 miles, until its speed at apogee is increased to 7000 mph.

Increase of speed is attained by fitting a rocket engine in the crate, usually called an *apogee boost motor* (ABM) if it is a solid-fuelled rocket, or a *liquid apogee engine* (LAE) if it is liquid-fuelled (Figure 16).

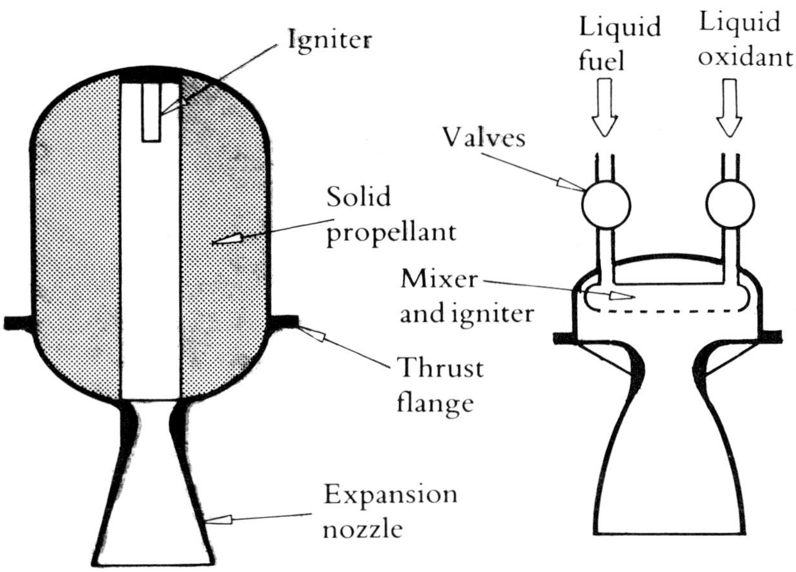

Figure 16 *Diagrammatic views of a solid propellant apogee motor (left) and a liquid propellant apogee engine (right)*

When Ariane is used for launching, it climbs on a curving path to about 200 miles altitude; it is then travelling at the required perigee speed of 22 800 mph parallel to the Earth's surface. At this point, it releases the satellite, which starts its journey on its elliptical transfer orbit, up to geosynchronous altitude and back again to perigee. When the satellite controllers on the ground are satisfied that the satellite is in proper working order (usually after two or three elliptical orbits) they command the satellite to slew into the right attitude (see Chapter 5) and then

Courtesy Arianespace

Launch of an Ariane rocket, carrying a British Aerospace ECS satellite into orbit

command the ABM or LAE to fire at apogee. The firing accelerates the satellite to the correct orbital speed of 7000 mph, and the orbit becomes circular (Figure 17).

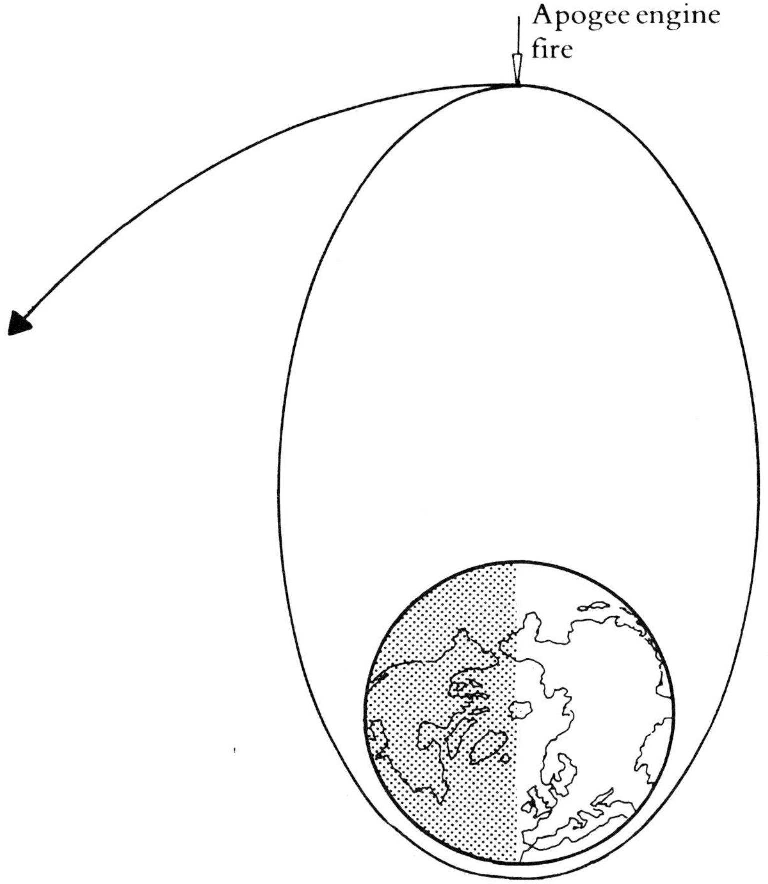

Figure 17 *Firing the apogee engine at apogee to increase the satellite speed from 3000 mph to 7000 mph puts it into a new circular path, the geosynchronous orbit*

The US Space Shuttle launch process is different. The Shuttle operates only in a circular orbit at about 150 to 200 miles altitude, and its speed at that height is constant at the balancing speed of 17 000 mph. With Shuttle, therefore, the satellite has first to be removed from the cargo bay, by means of either mechanical locks and springs or the Remote Manipulator Arm. It is then left for a short time in low circular orbit while the Shuttle pilot removes himself to a safe distance of several miles before any motor ignition occurs in the satellite. In this case, in addition to its ABM or LAE, the satellite is fitted with another rocket engine, usually called the *perigee assist module* (PAM), to increase its low-orbit speed from 17 000 mph to the required 22 800 mph (Figure 13). Once that speed has been attained, the satellite goes into its elliptical transfer orbit. Subsequent orbit circularisation is achieved in the same manner as with Ariane.

Another significant difference between the two launch methods arises from the different locations of the launch sites for Ariane and for Shuttle. Ariane is launched from Kourou in French Guiana, which lies on a latitude of about 5° North. Thus, when Ariane climbs to low circular orbit, the plane of that orbit will be about 5° off the equatorial plane that is required for the satellite. Shuttle is launched from Cape Kennedy in Florida, at a latitude of about 28° North. Its orbital plane is thus substantially more offset than that of Ariane (Figure 18). Plane-changing calls for the expenditure of propulsion fuel in the satellite: the more the offset, the more fuel is spent. This can have an adverse effect on the subsequent lifetime of the satellite, as seen in Chapter 5.

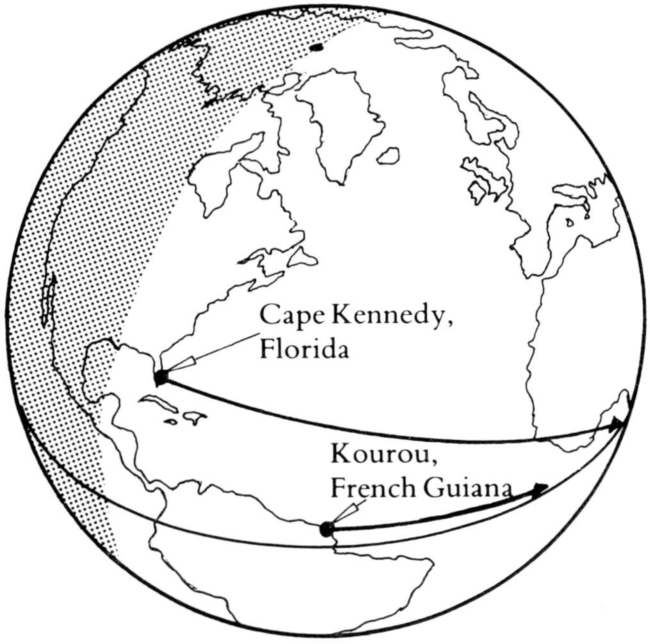

Figure 18 *A satellite launched by the US Shuttle from Cape Kennedy has to make more of a 'left turn' on to the Equatorial plane than does one launched by Ariane from Kourou. This uses extra on-board fuel*

After the Space Shuttle *Challenger* blew up in January 1986, the United States government decided that NASA, the agency responsible for the Shuttle fleet, should withdraw from launching any further commercial communications satellites, and that the fleet should be used only for launching military, scientific and other specialised payloads. As a result, more expendable launch vehicles (ELVs) are becoming available to compete with Ariane, such as the Martin Marietta Titan and the General Dynamics Atlas rockets. These also are launched from Cape Kennedy.

Note that, both at Cape Kennedy and at Kourou, launches take place eastward to take advantage of the initial speed imparted to the launcher by the Earth's own rotation. In both cases, too, the launch sites are located on eastern seaboards to ensure that spent rocket motors and other launcher debris fall into the sea.

4

Disturbances in orbit

The previous chapters may give the impression that all one has to do to put a crate into geosynchronous orbit is to accelerate it to 7000 mph at an altitude of 22 000 miles and send it on its way, after which it will go on circling the Earth in 24-hour synchronism for ever.

Alas, it is not that easy. If it were, the world's satellite manufacturers would have a very simple job and satellites would not cost what they do today. It is not easy because, out there at 22 000 miles altitude, the crate is subjected to a range of disturbances which, if left to do their worst, would inexorably take the crate out of orbit.

Disturbances, firstly, arise from the shape of the Earth itself. The Earth is not a sphere. It is flattened at the poles so that it resembles a giant squashed grapefruit (Figure 19) with the southern hemisphere larger than the northern. Because of this asymmetry, the centre of gravity to which the crate tries to fall is south of the line that we call the Equator. Worse, the centre of gravity is not in line with the spin axis of the Earth and this clearly puts the centre of the crate's 'falling circle' out of line with the Earth's centre. Additionally, the Earth is not even circular when viewed from the poles – it has bulges at the Equator which affect the position of what the crate 'thinks' is the Earth's centre of gravity. Finally, even the entire shape of the Earth does not stay constant – it vibrates like a gargantuan bag of jelly as its soft insides expand and contract.

The effects of all these movements and asymmetries culminate in distortions of the circular orbit and in rotation and tilt of the orbital plane itself. To the viewer on Earth, the crate will appear to move north or south from its intended position as the orbital plane tilts, east or west as the orbit rotates around the poles, and in complex other motions arising from distortions of the orbital circle itself.

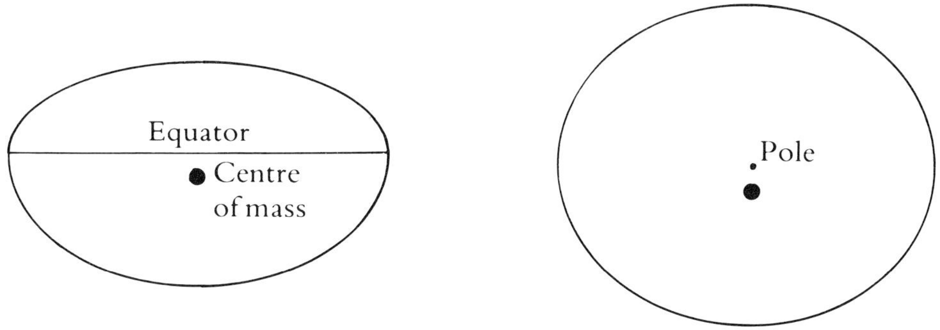

Figure 19 *The asymmetric Earth*

Even greater are the effects of the Moon. The gravitational attraction of the Moon on the Earth is seen in the tides, which show what is happening to water molecules in the sea as they are pulled by the Moon. Figure 20 shows why the pull gives rise to two tides per day. A is a molecule of water in the sea on the side of the Earth facing the Moon. It is 4000 miles nearer to the Moon than point B, which is the Earth's centre of gravity. C is another molecule of water on the other side of the Earth, another 4000 miles farther away from the Moon. So molecule A finds the Moon's attraction a little higher than does point B and thus moves away from B. In the same way, point B finds higher attraction than molecule C, which lags behind. Thus the seas are permanently distorted into an ellipsoid, as seen in Figure 20, with its major axis always pointing at the Moon, while the Earth rotates within it. In any 24 hours, therefore, a point on the Earth

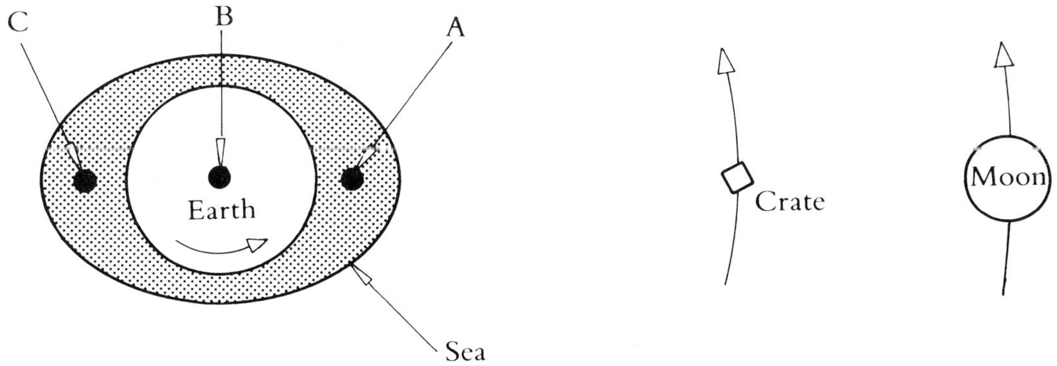

Figure 20 *Sea-water molecules* A *and* C*, and the Earth* B*, are all accelerating at different rates towards the Moon, because of their differing distances from it. This gives rise to two tides per day. The crate is subjected to similar 'tidal' pulls*

passes under two heaped-up sea masses, which cause our two tides per day. The fact that the Moon is also rotating round the Earth once a month means that the heaping is shifted, in the same direction as the Earth's rotation, about 12° per day. This puts about an hour's delay on tide peak, giving us our 13-hour tides.

If our crate is placed out in geosynchronous orbit, it too will be affected by the Moon in exactly the same way as the tides. It will move towards and away from the Earth in 13-hour cycles, and will appear to an observer on Earth to be shifting north and south.

There is also the effect of the Sun. At 93 million miles distance, its variation in gravitational attraction across the 50 000 mile diameter of a geosynchronous orbit is minimal. However, the Sun does exert a pressure on everything that falls within its rays. This 'solar wind', resulting from the heat and light energy radiated by the Sun, is both a problem and a help to the satellite designer. The pressure is tiny – about two millionths of a pound per square foot (or 0.5 kg per square kilometre) – but it is enough to push a satellite away, very slowly, to eternity. And yet, small as it is, this pressure can be used to control a satellite's attitude in orbit.

So, pushed by the Sun, pulled by the Moon, and falling round an oddly-shaped celestial body, our crate will certainly not stay in geosynchronous orbit. Left to itself, it will float away, very slowly, probably to one of the points between the Earth and the Moon's orbit where everything balances out to give no pull in any direction – zero-gravity points known as Lagrange or libration points. These are the junkyards of near space – they can be seen through a telescope because light is reflected off the clouds of galactic dust particles that have collected there over billions of years.

If the crate is going to stay in its right place in orbit, it has to be controlled. It has to carry equipment that allows it to be controlled. It has to be turned into a working satellite.

5

Turning a crate into a satellite

Think what the crate has to do. It has to stay in one place in its geosynchronous orbit so that it appears stationary to its users on Earth. It has to be lined up to the Earth so that one face of the crate – the 'talking–listening' face – is always pointing at the users down below. It has to generate its own electrical power to keep itself in order and to operate the communications equipment on board. It has to maintain itself at the right temperature inside: on the side facing the Sun, its skin may heat up to +150°C; on the opposite side, the skin encounters a temperature as low as −200°C. It has to operate in pure vacuum. And it has to go on operating, faultlessly, for up to ten years, with no one around to repair it if it goes wrong. Low-orbit satellites can, of course, be repaired, and have been so by astronauts working out of the Space Shuttle; but that is at an altitude of 200 miles, not where we are at 22 000.

Let us take these requirements, one by one, and see what has to go into the crate to handle them.

Staying in one place – station-keeping

The owner of a communications satellite usually guarantees to its users that the spacecraft will be kept on-station, at its assigned place in the geosynchronous orbit, with an accuracy of ±0.1°. This ensures that the users will never need to re-align their antennas on the ground.

At a range of 22 000 miles, this accuracy means that, to the user, the satellite will always be within a square of 40-mile sides (Figure 21). Its position within that square is monitored continuously by the owner, who uses a large tracking–ranging station for the purpose. When it is clear that the satellite is drifting towards one side of the square under the influence of all the orbital disturbances that it suffers, he takes action to bring it back.

Courtesy British Aerospace

The central structure of a British Aerospace Olympus *satellite being hoisted in the integration hall. Beneath it can be seen the apogee engine and station-keeping thrusters, and the spherical fuel-pressurisation tanks*

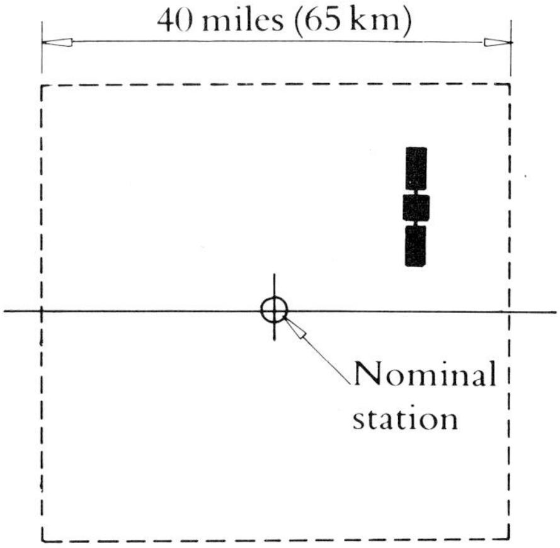

Figure 21 *The satellite must be maintained within a square of 40-mile sides centred on its allocated station in orbit*

The action consists of sending command signals to electrically-driven valves in the satellite which control the flow of propellant to very small rocket motors mounted on the structure. Typically, they provide a thrust of a few hundred grammes, which is quite enough to rotate and to move the spacecraft. Figure 22 shows how these rocket motors, or *thrusters* as they are called, can be used in pairs to drive the satellite in any direction or to rotate it. In the situation shown in Figure 21, where the spacecraft has to be moved back to the centre of the square, a combination of firings of pairs of thrusters will accomplish this task, under the control of a computer at the tracking station. At the end of the correcting movement, of course, other thrusters have to be fired to stop the satellite sliding out of the square on the other side!

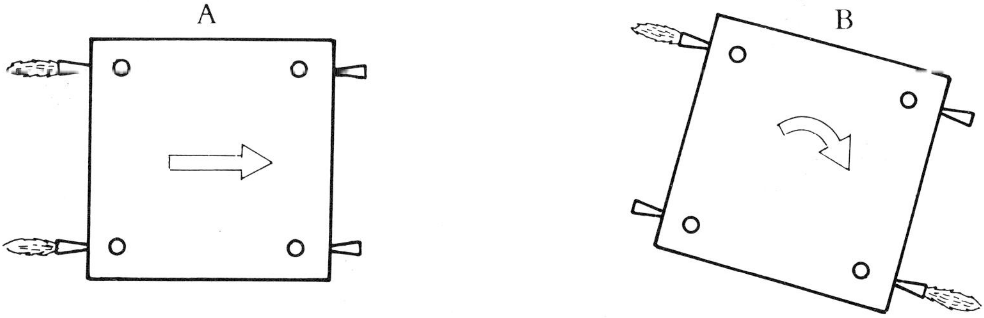

Figure 22 *Firing thrusters on one face of the spacecraft moves it in the opposing direction* (A). *Firing thrusters on opposite sides rotates it* (B)

Courtesy British Aerospace

An early mono-propellant thruster. The nozzle diameter is about 8 cm (3 in)

The propellant for the thrusters uses two liquids – monomethyl hydrazine as the fuel and nitrogen tetroxide as the oxidant – stored in separate titanium tanks and pressurised by helium stored in further tanks. Typically, a medium-sized communications satellite will use about 30 kg of propellant per year for correction movements. This fuel consumption determines the 'lifetime' of a satellite. The amount of fuel that can be stored on board is determined by the lifting capability of the launching rocket; the total mass of the satellite has to be controlled so that at least ten years' supply of fuel can be loaded. In addition, extra fuel has to be provided to fire the satellite out of orbit at the end of its life so that it does not block a valuable parking-space in the geosynchronous orbit.

The fuel and oxidant tanks in fact contain much more fuel than what is needed for station-keeping corrections. As explained in Chapter 3, the satellite must be accelerated to its correct orbital speed of 7000 mph at geosynchronous orbit from its transfer orbit apogee speed of 3000 mph. Most of the fuel on board is consumed by the liquid apogee engine that provides this acceleration. If a solid-fuelled apogee motor is employed, this extra fuel is not required. However, liquid engines provide more thrust per pound than solids, and offer an additional advantage: the amount of fuel put on board for the apogee engine is based on estimates that allow for errors in the positioning and speed of the satellite caused by the launcher; if those errors are less than estimated, less fuel is required, and the unspent fuel can be used to increase the lifetime of the satellite.

Our crate now contains a large rocket engine, a dozen or so thrusters, fuel tanks, oxidant tanks, pressurisation tanks, and all the piping and valving necessary to interconnect and control this equipment.

Courtesy British Aerospace

Integration of a Eurostar*-class satellite, showing two of the four fuel tanks and interconnecting pipework*

Lining up to the Earth – 'pointing'

If our crate is to be a communications satellite, it must carry antennas that will focus its signals on to specified areas on the Earth. The antennas must always point towards those areas and, because they are fixed to the satellite body, the body itself must ensure that one of its faces is always pointing towards the Earth. In effect, this means that the body must rotate in space once every 24 hours as it circles the Earth in orbit (Figure 23).

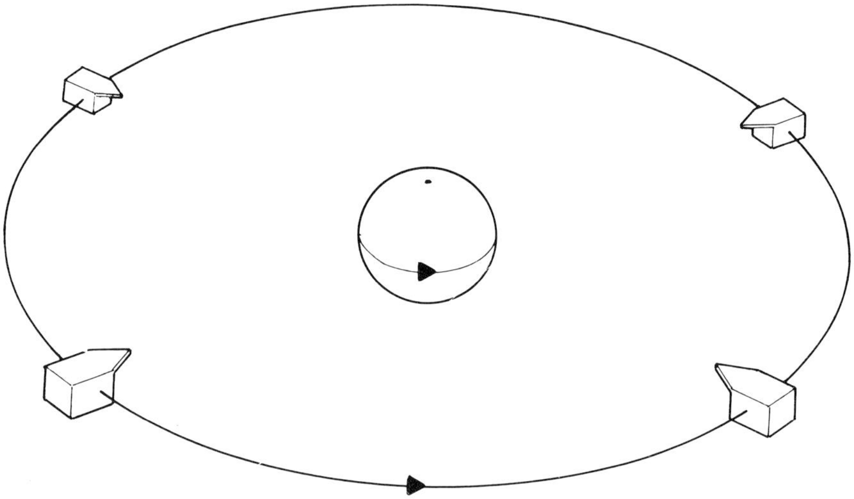

Figure 23 *While circling the Earth once every 24 hours, the spacecraft must also rotate once every 24 hours so that one face is always pointing towards the Earth*

This is easily achieved by commanding two thrusters to put the correct spin rate on the satellite at the beginning of its life in orbit. In theory, the body will go on rotating once every 24 hours for ever.

In fact, of course, this is not the case. Asymmetries in the shape of the satellite will cause solar wind pressure to rotate it, residual magnetic forces from the Earth's magnetic field will react with currents passing through wires in the satellite and generate rotational torques, and other asymmetries will give rise to gravitational effects that will attempt to re-align the body. All of these effects could be overcome by employing the thrusters to compensate body movements – but the thrusters use valuable fuel. What are used instead are devices called *momentum wheels* and *reaction wheels*, which use only electrical power.

The action of a momentum wheel is shown in Figure 24. It is an electrically-driven flywheel, typically weighing about 2 to 3 kg, and spinning at around 2000 rpm, housed in a vacuum-tight container. It is mounted so that its spin axis is aligned north–south, on the 'pitch' axis of the body. In this position, it acts as a gyroscope, giving inertial stiffness to the body in the other two axes ('roll' and 'yaw') while permitting the body to rotate around it in pitch once every 24 hours. Now the body is 'stiff' in roll and yaw, and it can be controlled in pitch simply by speeding up or slowing down the wheel.

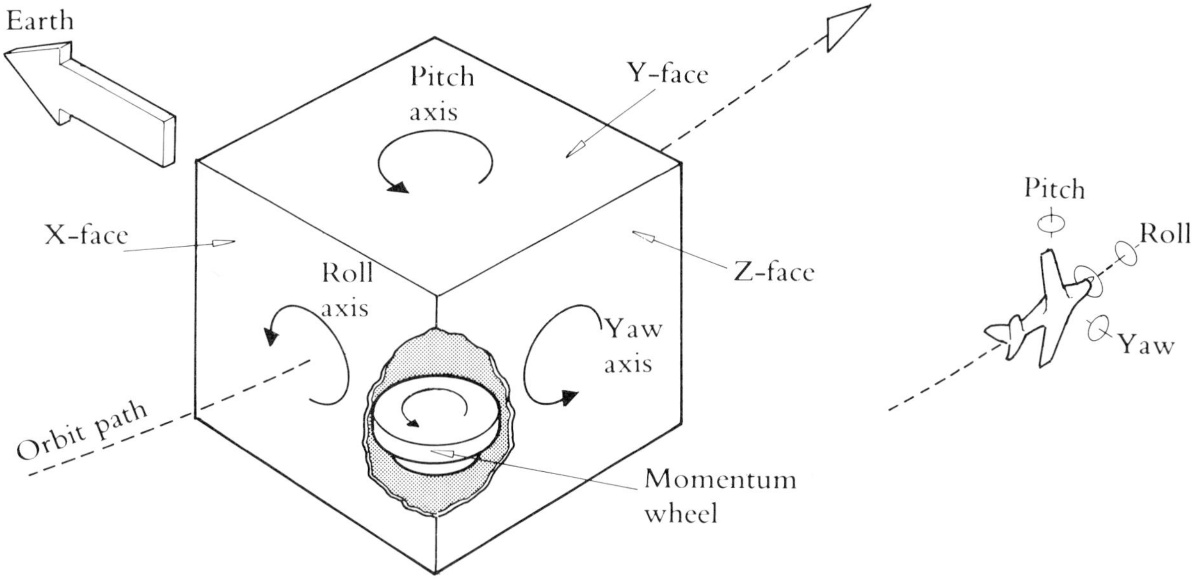

Figure 24 *Installation of a momentum wheel inside a spacecraft body. Changes in wheel speed cause rotational movement on the pitch axis. The sketch of the aircraft shows how the names of the three spacecraft axes have been adopted from aeronautics. The position and designation of the body faces are also shown*

Suppose, for example, as in Figure 24, that the body is tending to swing in pitch anticlockwise. This would mean that its antenna beams would be swinging westward on the Earth. To compensate that movement (which is recognised by sensors on the satellite that watch the infra-red radiation of the Earth's disk, or by other sensors that are locked on to radio transmissions from the ground) the wheel is commanded by an on-board computer to slow down by a few rpm. This change of momentum in the wheel has to be compensated by the body, which swings a fraction of a degree clockwise, so that its beams shift eastward again. Speeding up the wheel will move the body in the opposite direction. When wheel-speed changes are completed, and the wheel is running at constant speed, the body feels no changes in momentum and remains pointing at the right place on Earth. Eventually, if the body has been continuously twisted in one direction by solar pressure or other disturbances, the repeated speeding-up or slowing-down of the wheel to correct the movements will take the wheel to its maximum safe speed, or slow it down to the extent that it can no longer act as a gyroscope. At this point, the wheel must be returned to its median speed. To overcome the violent reaction of the body to these big changes in momentum, thrusters are provided to react against the change and to 'dump' the momentum.

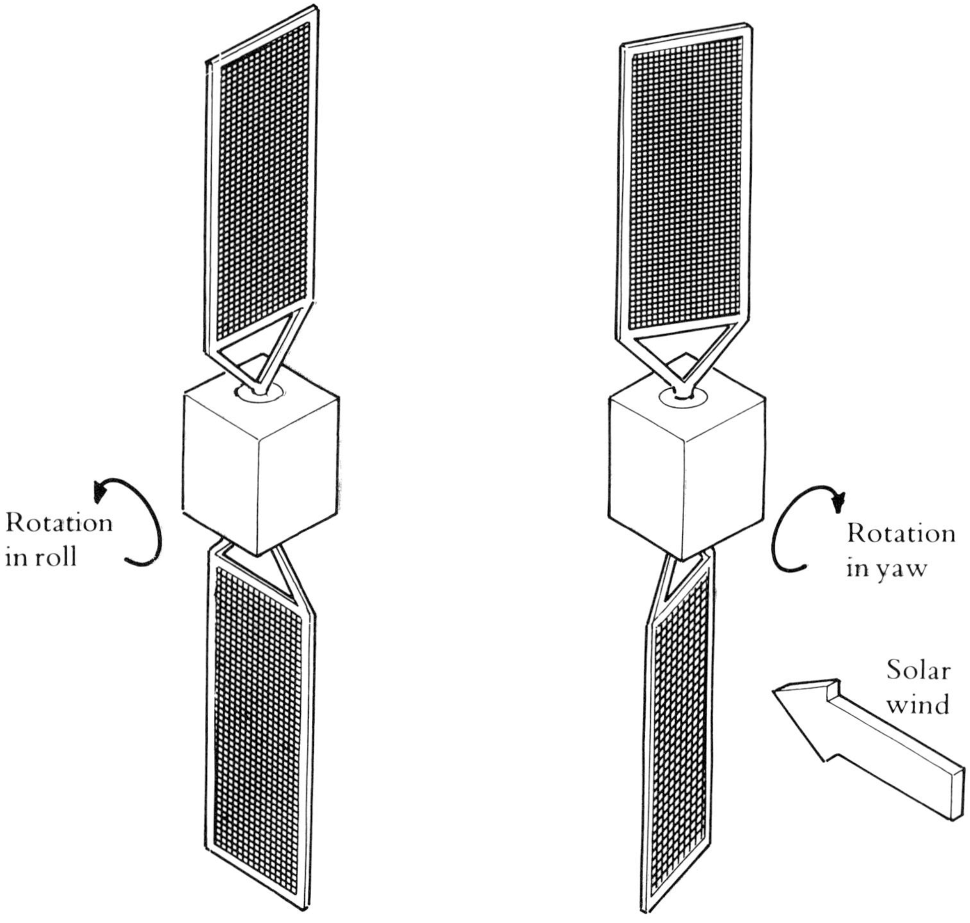

Figure 25 *Rotating the arrays off the Sun-line by ground command or by on-board control provides a means of rotating the body in roll and yaw. The off-line angles shown here are exaggerated for clarity. To minimise these angles to conserve solar power, angled 'flaps' are sometimes fitted to the tips of the arrays*

The 'stiffness' of the body in the other two axes, roll and yaw, arising from the unwillingness of a gyroscope to let its spinning axis be moved, gives adequate pointing control of these axes, but they too must be capable of correction from time to time. Excessive rotation in roll will cause the satellite's beams to swing north and south, while movements in yaw will cause the beams to rotate. With a satellite carrying a momentum wheel, which corrects pitch movements only, two methods are used to control the other two axes. One method is to employ two momentum wheels, set at a slight angle to each other, but with their 'average' axis still set on pitch. Run as a pair at equal speeds, they will act as a double-sized single wheel; if the speed of one is changed, it sets up a momentum change at an angle to the main axes of the body, all of which react accordingly. Thus, pitch, roll and yaw axes of the body can be changed together, using an on-board computer to handle the complex dynamics equations that are involved. Another method is to use the solar arrays (see the next section) as sails in the solar wind. By being turned at angles off the Sun, they can generate enough torque to rotate the body in roll and yaw (see Figure 25).

Momentum wheels are simple and effective. One class of satellites, known as *spinners*, use their own bodies, which rotate permanently at around 40–60 rpm, as momentum wheels (see Figure 26).

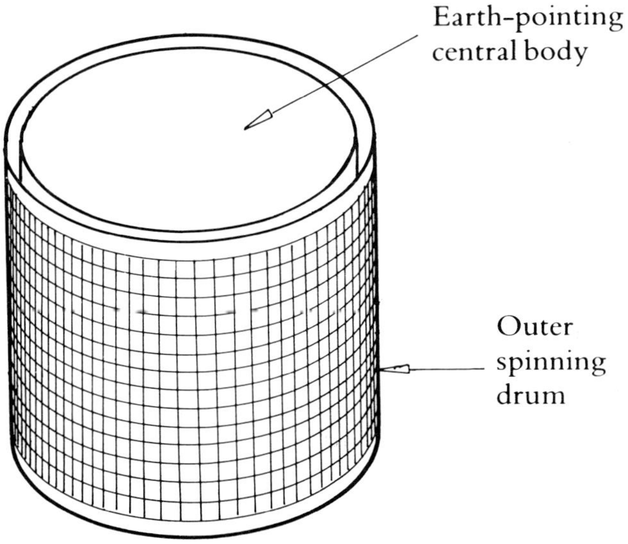

Figure 26 *A 'spinning' satellite in which the outer drum, which carries solar cells for power, also acts as a momentum wheel*

Courtesy British Aerospace

Momentum wheel and wheel-drive electronics

Momentum wheels have their limitations, however. They are designed to accommodate a fixed range of inertias that are expected to be found in particular classes of satellites. If this range is exceeded, they become less effective. The large satellites now coming into operation can carry many different forms and masses of electronic equipment, antennas and solar arrays, with widely differing rotational inertias. For these satellites a different pointing control system is employed, using reaction wheels.

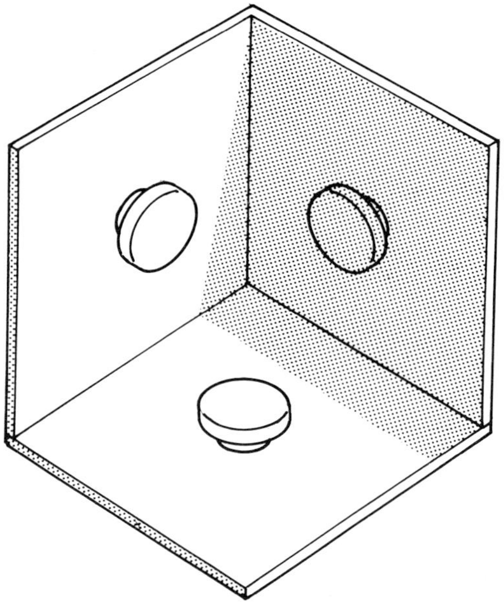

Figure 27 *Installation of three reaction wheels inside a spacecraft body. Change of speed of any wheel causes rotational movement of the body on that axis*

Reaction wheels (Figure 27) are smaller versions of momentum wheels, mounted differently. They too are small, electrically-driven flywheels housed in vacuum-tight containers, but they are used in threes, one wheel in line with each body axis. Each acts on its own axis in exactly the same manner as does a momentum wheel on the pitch axis: if a wheel is accelerated, the body will react by turning slightly in the direction opposite to the wheel's spin; if a wheel is decelerated, the body will shift slightly in the same direction as the wheel's spin. Like the momentum wheel, their speeds and directions are controlled by an on-board computer that takes its information from infra-red sensors looking at the Earth, from inertial gyroscopes on board, and from ground commands. Usually, a fourth wheel is fitted, mounted at an angle to all the rest, which can take over in the event of a failure of any of the three main wheels, the effects of its angle being compensated by the on-board computer.

One other additional method is used for pointing. Called a *magneto-torquer*, it comprises a loop of wire set in the body. An electrical supply fed into the loop causes it to react with the Earth's residual magnetic field and turn the body in the same way as do armature wires in an electric motor. Very small and precise correction movements can be made with this device.

Our crate is being filled with more equipment. It now contains a momentum wheel or reaction wheels, infra-red earth sensors, inertial gyroscopes, a magneto-torquer, and an on-board computer to control the entire pointing system. Also it already needs a telecommand receiver to accept control commands from Earth, and a telemetry transmitter to tell those down at the tracking station what is going on inside the body. The need for electrical power is obviously urgent.

Electrical power

All communications satellites in operation today use solar cells to generate their electrical power. Other forms of power generation in space do exist, such as nuclear reactors, but none of them can offer the simplicity, safety, reliability and economy of solar cells.

A solar cell consists of two layers of silicon, differently 'doped' so that they offer different resistance to the passage of electrons through them. The outer – light-collecting – layer is so thin that it is transparent to light. When light from the Sun hits the cell, the light energy, in the form of photons ('packets' of energy), passes through the outer layer and displaces electrons within it. Because of the doping of the silicon, the electrons can pass only into the back layer of the cell, where they are collected and fed into wires. This flow of electrons through the wires is the generated electrical power required by the satellite. Later, when the power has been passed through the electrical equipment on board, the electrons are returned to the front faces of the cells through metal connections on those faces.

The light-to-electricity conversion efficiency of solar cells is very low – about 10% – but because the prime energy is cost-free, this is of little consequence. The only adverse effect is that the remainder of the energy – 90% – appears as heat, which has to be radiated out into space through the backs of the cells. Out in space, the Sun's radiation provides 1400 watts of energy per square metre, so an array of solar cells will provide 140 watts per square metre. This 140 W/m^2 is available only if the cells are face-on to the Sun. If they are angled away from it, their power output drops until, at a 90° angle, the power output is zero. This raises the need for another piece of equipment to be fitted in our crate.

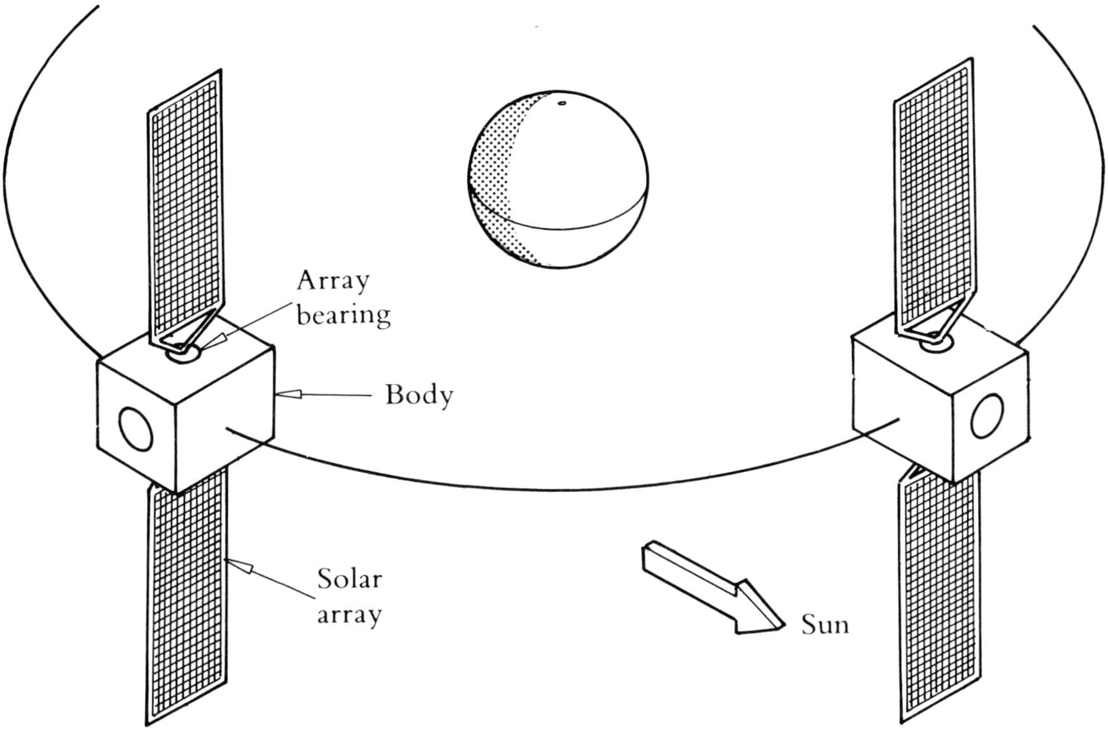

Figure 28 *A solar array rotational bearing, that rotates once every 24 hours, is needed to keep the arrays pointed at the Sun while the body points at the Earth*

Figure 28 repeats Figure 23 in that it shows the crate turning once every 24 hours so that one face is always pointing to the Earth. But Figure 28 shows the direction of the Sun too, and the Sun is so far away that it is effectively at a fixed point in space at any given time. So, while the crate is rotating, the solar cells that it carries must face that fixed point all the time if they are to generate their maximum output. Figure 28 shows how this is done. The solar cells are mounted on panels, called *solar arrays*, which in turn are mounted on rotating bearings that are fixed to the satellite body on its 'north' and 'south' faces. The bearings are driven by electric motors which are commanded by sensors on the arrays that lock on to the Sun.

Courtesy Matra SA

Assembly of solar cell arrays on to an ECS-2 satellite

Thus, as the body rotates in space once per day, the bearings also rotate in the opposite direction, at the same speed, holding the arrays face-on to the Sun. The arrays can also be offset from the Sun-line to apply rotational torque to the satellite body (Figure 25).

The bearings are built under meticulous conditions of cleanliness, because they must maintain their once-per-day rotation for ten years without failure and without lubrication (for lubricants evaporate out into the vacuum of space). In addition, they have to pass the electrical power generated by the arrays, which may amount to 8 kilowatts or more, into the satellite body, and they therefore carry multiple electrical slip rings, which also have to run dry.

Spinning satellites, which use their own bodies as momentum wheels, do not have to carry these bearings. Their cylindrical bodies are covered with solar cells, some of which are always looking at the Sun (Figure 26).

Courtesy Matra SA

A Telecom 1 *satellite in integration, with one solar array fully extended, supported by an overhead gantry*

Because only one line of cells is directly facing the Sun at any one time, while all the others are angled to it, cylindrical satellites have to carry over three times as many cells as the equivalent flat panel of the same projected area. In addition, they suffer from the disadvantage that the solar cell output is limited by the surface area of the body – unlike our 3-axis stabilised crate, which can carry flat solar-cell panels of almost any size.

Solar cells last for ever – many test arrays on Earth have given power for twenty years or more – but out in space they are subjected to micrometeorite and galactic dust bombardment, which slowly reduces their output, typically at a rate of 2% to 3% per year. Because the satellite needs a specified amount of power right up to the end of its ten-year working life, the satellite manufacturer puts additional cells on the array to compensate for this loss. This means that, at the beginning of the satellite's life, the arrays produce some 30% or more excess power. The excess has to be radiated out into space as heat through further devices called *shunt dumpers*, which in turn are controlled by a central voltage and current regulating system.

The electrical power system is not yet complete. It needs even more equipment because, twice a year, the satellite passes into eclipse. Figure 29 shows why, and when, eclipsing of the satellite occurs in the geosynchronous orbit. The Earth's spin axis is inclined at 23° to the plane of the Earth's own orbit round the Sun (the ecliptic plane) so the Earth's Equator is inclined at that angle too. And, because it is in line with the Equator, so is the plane of the geosynchronous orbit. Looking first at midsummer in the northern hemisphere (to the right of Figure 29) it can be seen that the satellite will be in sunlight around its entire orbit – even when it is 'behind' the Earth, it can see 'over' the top of the Earth's shadow. A similar situation occurs in midwinter, when it can see 'under' the shadow. But at equinox (March 21st and September 21st) the satellite passes into the Earth's shadow for a period.

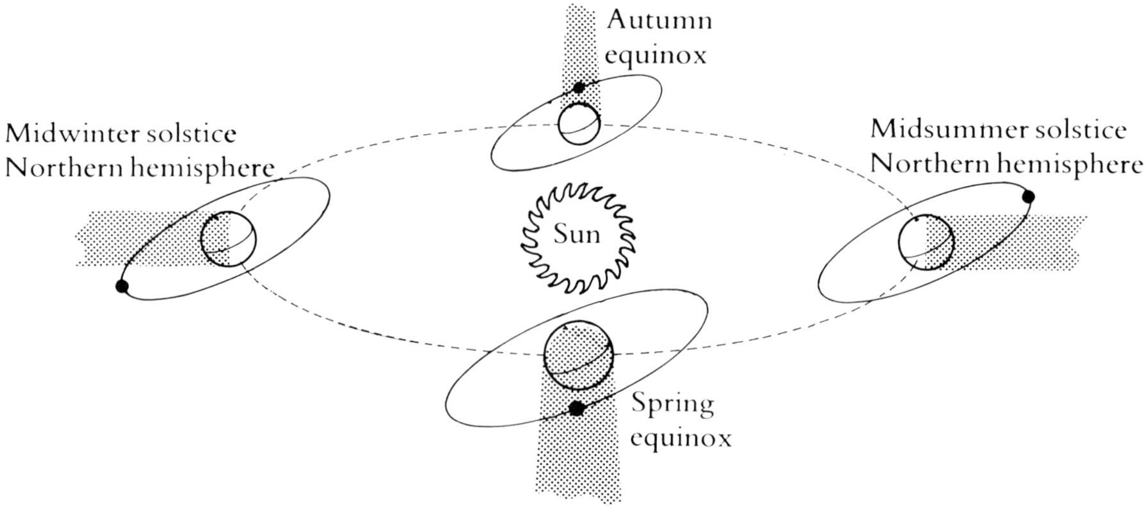

Figure 29 *The geometry of geosynchronous eclipse. At solstice, the satellite can 'see' the Sun throughout the 24 hours orbit time. At equinox, the satellite passes through the Earth's shadow*

This shadowing does not occur suddenly at equinox. The satellite starts to clip the shadow some twenty days before each equinox, and will not be totally clear of it until twenty days after. The worst case occurs at the equinox dates, when the satellite has to traverse the entire width of the

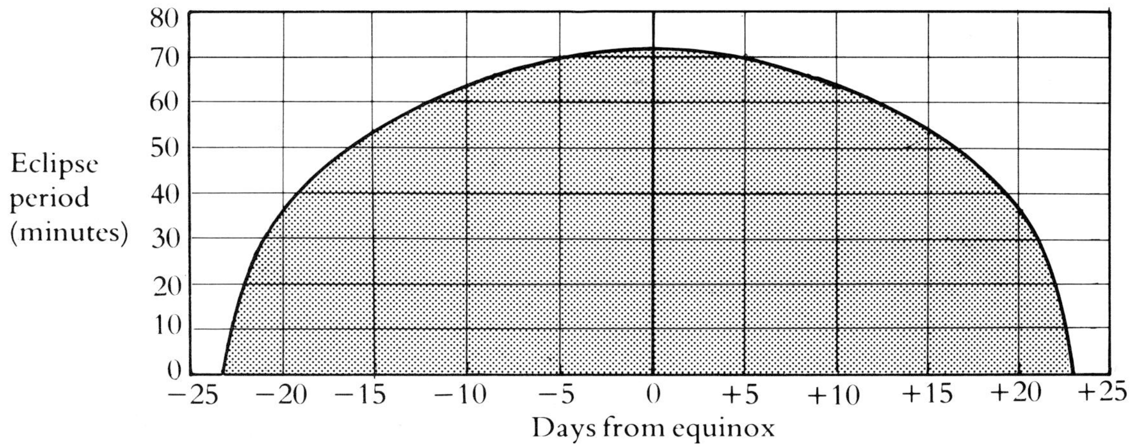

Figure 30 *Duration of eclipse periods before and after equinox*

shadow. The width of the shadow is the diameter of the Earth – about 8000 miles – so the satellite, travelling at 7000 mph, will take about 70 minutes to pass through it. The pattern of increasing eclipse time up to equinox, and its decrease again after equinox, is shown in Figure 30.

In eclipse periods, the solar cells will not generate power, but the satellite still needs power, and this is provided by batteries. Satellite batteries use lightweight, highly-efficient nickel–cadmium or nickel–hydrogen cells that offer up to 2 ampere-hours per kilogram for 10 years or more. The size, and thus, more importantly, the mass of the battery is determined by how much power the satellite will need in eclipse, which need not be the same as that required in sunlit hours.

To an extent, battery power requirement is a matter of the timing of the 70-minute eclipse period during the user's night. Figure 31(a) shows another view of the geometry of the eclipse at equinox, with the satellite stationed on the same longitude in orbit as the user's longitude on Earth. If the user were in London, for instance, at 0° longitude, the satellite would be stationed over the Equator also at 0°. In this case, the user will see the satellite enter eclipse at about 11.25 p.m., about 35 minutes before it passes his midnight position on Earth. Seventy minutes later, at about 12.35 a.m., the satellite will come out of eclipse. Loss of power at this time of the late evening is unacceptable to television companies and to telecommunications authorities. Both would expect full power capability to continue, and thus the satellite would have to carry batteries that supplied the same power as that coming from the entire solar arrays.

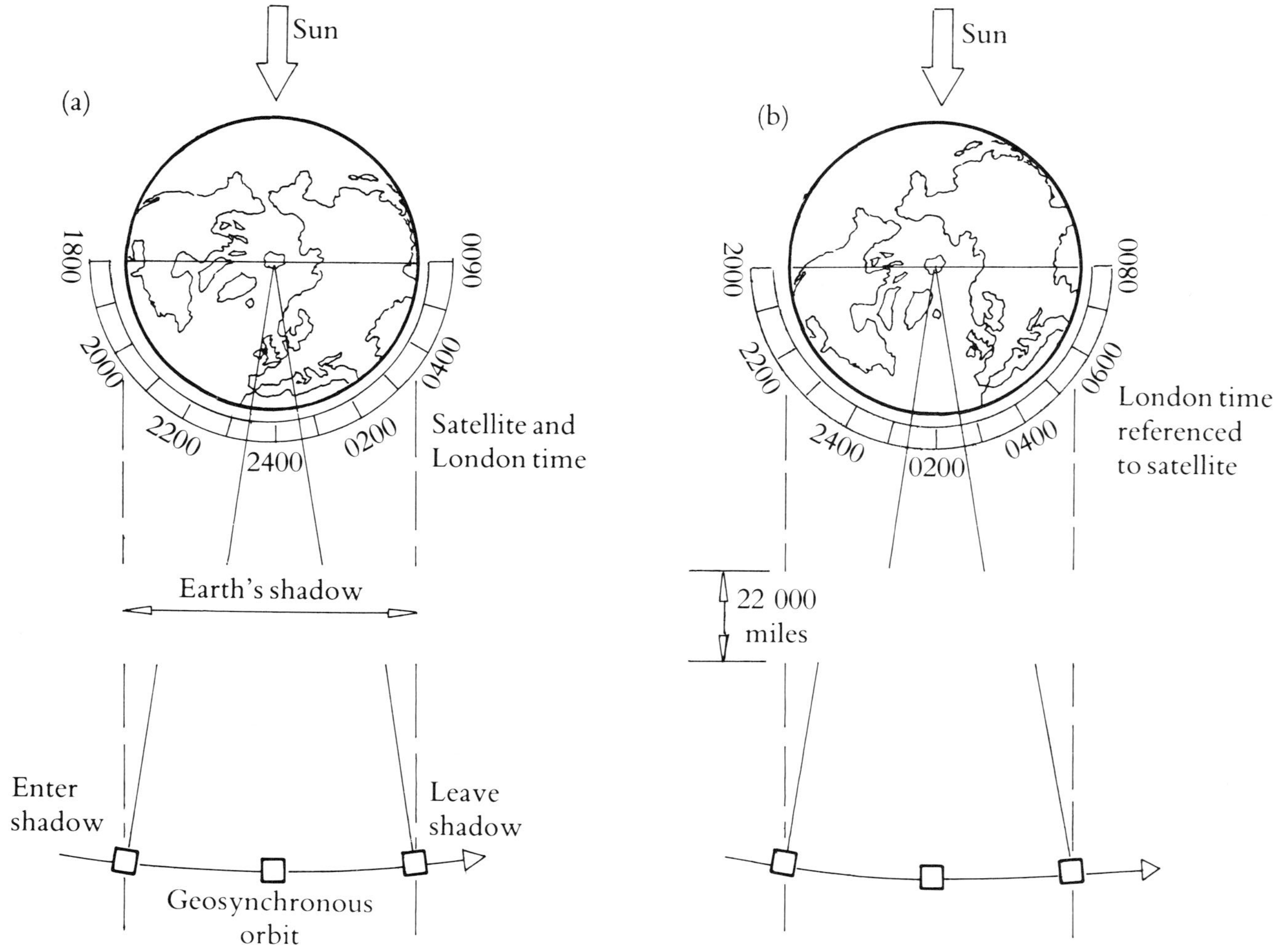

Figure 31 *Locating a satellite westward of the service area (London) to delay eclipse.* (a) *The satellite is at 0° longitude, and enters eclipse at 11.25 p.m.* (b) *The satellite is at longitude 30° West, and enters its eclipse at 1.25 a.m. London time*

Battery power provision can be reduced by moving the satellite in relation to the users. Figure 31(b) shows the same geometry, but now the satellite has been moved to new station, 30° West, over the Atlantic, while the users remain in London. Now, as the geometry shows, the users see the satellite enter eclipse at around 1.25 a.m. and come out again at 2.35 a.m. This is a much more convenient time for the TV people to accept an outage, while even the telecommunications authorities, who normally aim at 24-hour operation, will accept that their traffic is lower at that time of the morning. So, for TV, there may be no need for any batteries at all during eclipse, while some telephone authorities will accept a power reduction of 40% to 50% for that 70 minutes. The resulting substantial saving in battery mass leaves more mass allowance in the satellite for station-keeping fuel, and thus a longer lifetime. This is why most communications satellites are stationed to the west of their service areas.

The only requirement for full 24-hour operation comes from the satellite itself, which has to keep all its own equipment, described in the earlier parts of this chapter, in operation all the time. Typically, this consumes between 250 and 400 watts continuously, which means, with a 50-volt operating system, between 5 and 8 amps over the eclipse period, and thus a battery capacity of 10 to 16 Ah, allowing for 60% charging–discharging efficiency.

Now our crate is nearly complete. It needs only one more set of devices – those that will control its temperature.

Thermal control

As we saw at the beginning of this chapter, the sides of the crate facing the Sun might heat up to +150°C if left without protection, while those looking out into space will encounter temperatures as low as −200°C. Figure 27 shows that four faces will be looking at the Sun at some time during each orbit – the two that always point directly at the Earth and away from it, known as the Z-faces, and the other two that always face east and west – the X-faces. The final two, those that carry the bearings for the solar arrays, and that look nominally north and south, called the Y-faces, do not look directly at the Sun at any time but, as Figure 29 shows, they do 'see' the Sun at a low angle of 23° at midsummer and midwinter – the solstices – and at lower angles for the rest of the year. At equinox, they do not see the Sun at all.

Inside the crate there is now a host of electronic, electrical and mechanical devices that will work happily only if their temperature is kept similar to that at which they were built and tested on Earth – about 20°C. Heat coming in from the X-faces and Z-faces, and even through the angled Y-faces, has to be blocked off, while heat from inside, generated by the electronic equipment itself, has to be let out in a controlled manner.

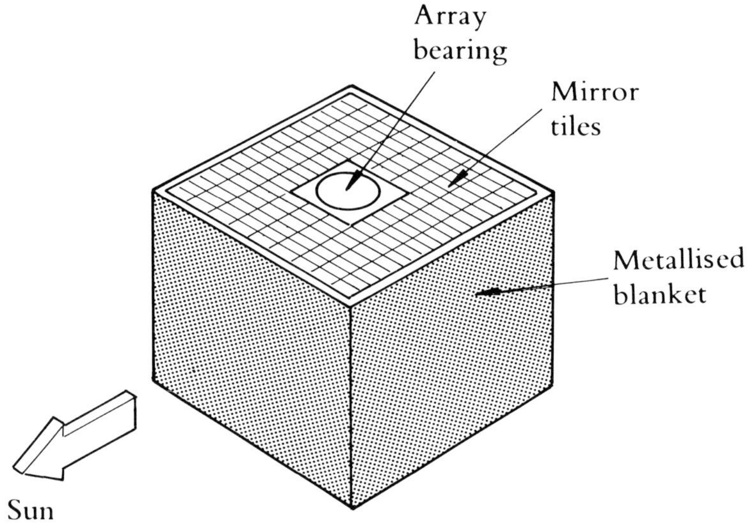

Figure 32 *Thermal control surface. The Y-faces, which 'see' the Sun only at a low angle at solstice, are covered with mirror tiles that reflect the Sun's heat and allow internally-generated heat to escape. The remaining faces are covered with metallised blankets that reflect direct heating from the Sun while retaining internal heat*

Two control methods are used – passive and active. In a passive system (Figure 32) the X-faces and Z-faces are covered with metallised thermal blankets. The high-gloss outer metallic finish of the blankets reflects most of the heat from the Sun, while the thermal properties of the blankets themselves prevent heat escaping through those faces. Heat is allowed to escape through the other two Y-faces – those carrying the solar array bearings – which are not subject to continuous heating variation from full Sun to the full cold of space. They are covered with mirror tiles, which reflect heat trying to enter the crate but which allow internally-generated heat to pass out through them. Tiles, each a few centimetres square, are used because it is easier and cheaper to replace a few tiles if the face is damaged during assembly than to change a large mirror covering the whole Y-face.

A passive system has to be supplemented by internal heaters that are switched on by ground command and supplied from the batteries during eclipses, because in those periods no heat is received from the Sun, and the equipment will run too cold without help from the heaters. Heaters are also used when parts of the satellite are shut down, for example when a communications channel is taken out of service temporarily, to compensate for the waste heat normally generated by that channel's equipment.

Active systems are added for areas of very high heat generation such as high-wattage amplifiers and their associated output filters, which generate a lot of waste heat over a small mounting area. The transfer of heat away from these areas is accomplished by 'heat pipes' – loops of tube filled with refrigerant. The 'weightlessness' of space is exploited here. The refrigerant fluid is pulled along the pipes by osmotic action on fabric or plastic 'wicks', collecting heat and increasing its vapour pressure as it goes, until it dumps the heat elsewhere and is returned to the beginning of the loop by the pressure differences in the closed system. In the gravitational conditions on Earth, the osmotic action would work over only very small distances; the refrigerant would have to be pumped, as in a domestic refrigerator.

It is the matter of waste heat radiation that determines the amount of power that can be fed into a satellite from its solar arrays, and the waste heat radiation is in turn dependent on the size of the satellite itself. We have seen that waste heat is radiated through the mirrors on the Y-faces, but this radiation is limited. At the working temperatures met inside the body, the Y-faces will radiate into cold space at a maximum rate of about 500 watts per square metre. The Y-faces of many of today's communications satellites measure about $1\frac{1}{2}$ metres square – a size determined by the launching rockets of the 1970s. This provides an area of just over 2 square metres. Thus, the 4 square metres of the two Y-faces will dissipate about 2 kilowatts of waste heat. The waste heat has been generated by electronic equipment inside the satellite, and that equipment wastes about two-thirds of the electrical power put into it. So power to be fed into the satellite from the solar arrays is restricted to 3 kilowatts, because two-thirds of it is going to be wasted as heat and 2 kilowatts is the maximum that the Y-faces can handle.

It is the larger launchers of the 1980s – Ariane and Shuttle – that now allow larger spacecraft to be launched. Because they are physically larger, with more area of Y-face, they can accept more power from bigger solar arrays – up to 8 kilowatts input – and thus offer more communications capacity from a single parking-space in the geosynchronous arc.

Nearly a working satellite

Now our crate is a working satellite – or nearly so. Fitted with all the equipment described in this chapter, and filled with fuel, it will accelerate itself to the right orbital speed after it leaves its launching rocket, it will accept commands from Earth to keep itself exactly on station in orbit, it will ensure that one of its faces is pointing towards the Earth at all times, it will generate its own electrical power in sunlight and in eclipse, and it will keep itself at the right temperature. And it will do all this for ten years.

But that is all that it will do, which is no use to anyone on Earth. It is certainly an Earth satellite, but it is not yet a communications satellite. To be useful in that role, it needs more equipment – the *communications payload.*

6

On-board communications equipment

To the user, a communications satellite is simply a microwave repeater mounted on a mast 22 000 miles high. And that, to the user, is all it should be. Before the advent of space communications, he had to send his communications and television via microwave repeater stations mounted on hilltops and mountaintops, each with a range just sufficient to reach the next repeater town about 30 miles away. So, to carry signals 300 miles from London to Edinburgh, he would need (and still does) 10 microwave stations, each with its mast, receiving and transmitting antennas, receivers and amplifiers, and local power supplies taken off the national grid or supplied from a diesel generator. To carry a TV programme from Athens to London, hundreds of towers would be needed on a route that might pass through six countries, with each national telecommunications authority applying its carrying charge on the way.

With a communications satellite, all that is needed is a transmitting station, one repeater on board the satellite, and a receiving station that may be up to 5000 miles away from the transmitter, or as near as 5 miles. In practice, satellites carry many repeaters, or *transponders* as they are generally called. Today, provision of 24 transponders is common, while new, larger satellites now in construction will carry 50 or more.

Transponders

A typical transponder is shown in Figure 33. The receiver section contains a pre-amplifier, filtered and tuned to accept the frequency band allocated to the service it is handling (see Chapter 8), and a down-converter, which converts the frequency of the up-link signal to the allocated down-link frequency back to the ground. Now at its new frequency, the signal is passed into a high-power amplifier, and then into an output filter.

It is the high-power amplifier that determines the performance of the satellite as seen by the user on the ground. Over a given service area on Earth, the higher the power output of the amplifier at the down-link radio frequency (RF output) the higher the received power on the ground, and the better the quality of the signal seen or heard through a particular ground station. RF outputs vary from 10 watts to a present maximum of 400 watts per transponder, depending on frequencies used. At the lower end of the frequency bands used for space communications, the amplifiers may be of solid-state construction, that is, transistor amplifiers. At the upper end of the range, travelling wave tubes (TWTs) are used, in which the incoming signal is fired along a tube that is carrying an electron beam which amplifies the signal as it traverses the length of the tube.

As we have seen, the higher the RF power output, the better the received signal, but there are limits to the level of output power that can be applied in a particular satellite. We have already noted the relatively low efficiency of high-power amplifiers, which means that only one-third of their input power appears as RF output power while the other two-thirds appear as waste heat, and that the amount of waste heat that can be dissipated into space is dependent on the physical size of the satellite. It is clear that, if that satellite has to carry a certain number of amplifiers to handle a specified amount of traffic, the RF output powers of those amplifiers must be restricted. Alternatively, if the RF output power is specified, then the number of amplifiers is restricted.

For example, take a satellite body that can dissipate 2 kilowatts of waste heat. This means that its solar arrays can deliver a maximum power of 3 kilowatts into the body, with two-thirds coming out as heat and one-third remaining available for RF power. So the total RF output of the satellite can be 1000 watts, which means that it can carry a variety of different 'payloads' of differing numbers and powers of transponders:

4 × 250 watts
or 10 × 100 watts
or 20 × 50 watts
or 50 × 20 watts

In theory, it could also carry 100 × 10 watts, but it is unlikely that that number of low-power transponders could be physically accommodated inside the body.

The final choice of numbers of transponders and their powers, and their relationships with traffic density and service areas, are dealt with in more detail in Chapter 11.

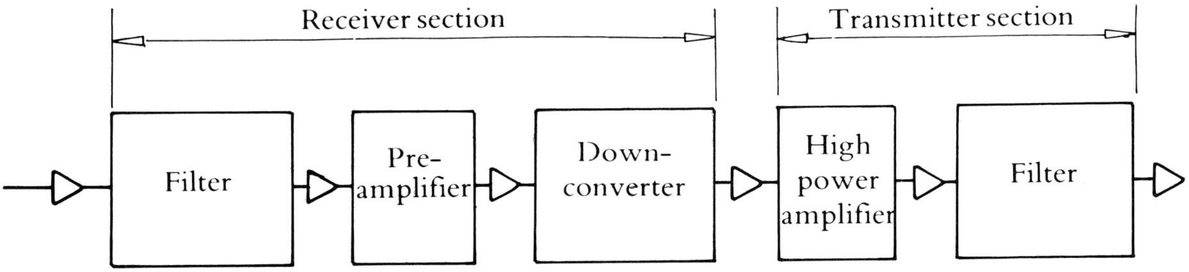

Figure 33 *Principal components of a transponder*

When numbers of transponders are carried in a satellite, they are invariably connected together through switching systems on board. Figure 34 shows an interconnected six-transponder payload, in which incoming signals may be routed to selected down-converters and TWTs for subsequent transmission. The routing can be accomplished by ground

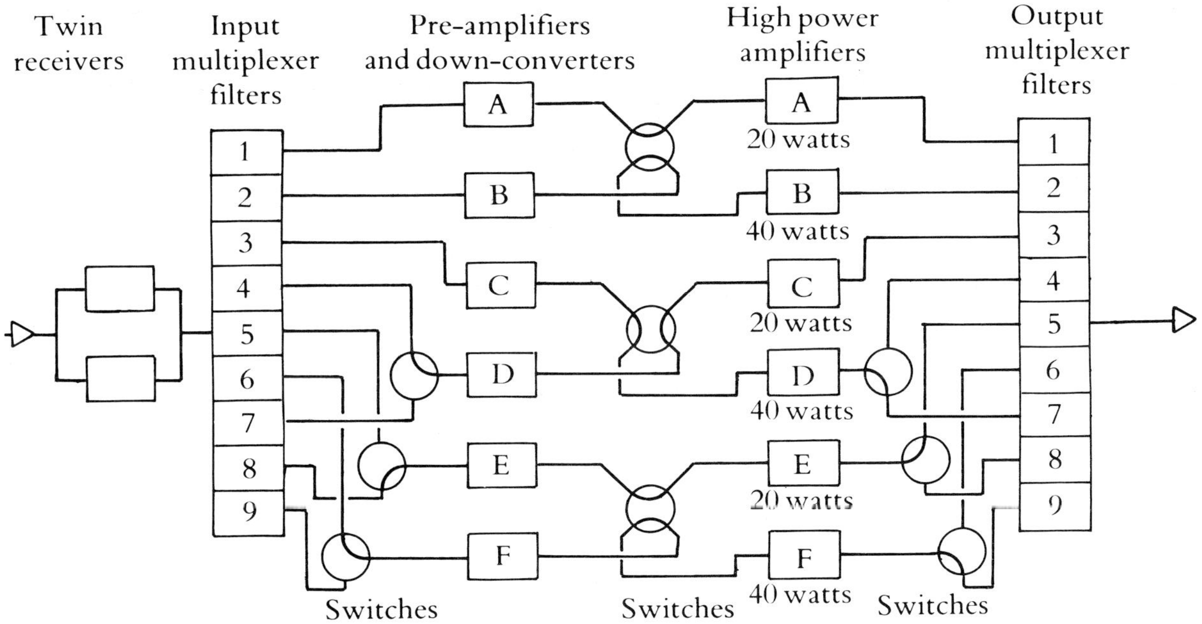

Figure 34 *An example of a switched 6-transponder block in a 24-transponder satellite. In this example, input channels* 1 *and* 2 *are 'hard-wired' to outputs* 1 *and* 2, *but may be switched between high-power amplifiers* A *and* B, *which offer different output powers. The remaining input channels may be switched to a variety of outputs as indicated. The two-way switches are shown in their alternative positions so that other switching routes may be traced*

command or by on-board controllers which recognise destination addresses in incoming signals and direct them to the appropriate output.

Additional systems are always included to bring standby equipment into operation in the event of failure. As in all terrestrial telecommunications systems, all receivers, pre-amplifiers and converters are duplicated, and in many cases the high-power amplifiers are fully duplicated too. However, the size and mass of the top range of high-power TWTs usually precludes full duplication, and it is usual to provide one standby for each pair of operating TWTs (known as '3-for-2 redundancy') or for each group of three TWTs (known as '4-for-3 redundancy').

Facilities to switch incoming signals and to re-route them on board are becoming increasingly important as satellites are required to handle complex traffic patterns covering a variety of service areas. The service areas, or *coverage areas* as they are often called, are dependent on another major component of the on-board communications equipment – the satellite antennas.

Antennas

After final high-power amplification and filtering in the satellite transponders, radio signals destined for transmission to Earth are passed into waveguides – tuned rectangular-section tubes that carry the RF energy to the outside of the satellite. If the waveguides were simply sawn off as they protruded from the satellite wall, the RF energy would radiate out into

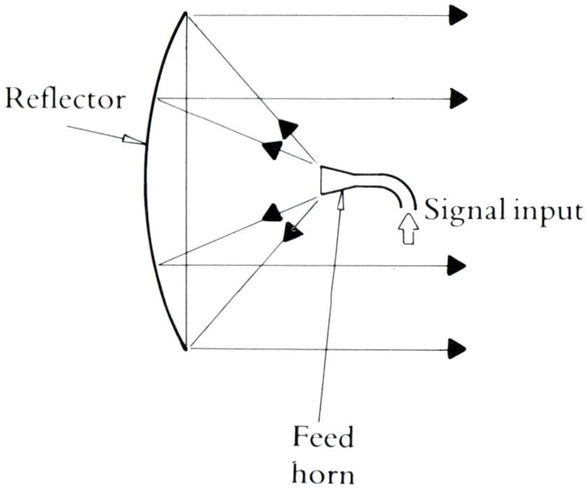

Figure 35 *Geometry of a centre-fed antenna*

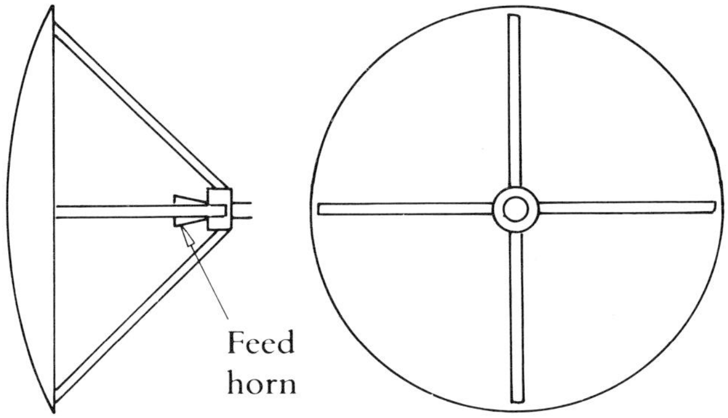

Figure 36 *A centre-fed antenna, showing aperture blocking by the feed horn support struts*

space in ever-expanding spherical waves (known as isotropic radiation) and only a minute fraction of the energy would reach the Earth. The task of the antennas is to capture that radiation as it leaves the satellite and to focus it on to specified areas on the Earth.

The most common type of antenna is a simple parabolic reflector, very similar to those used in optical systems for focusing light. The RF energy is carried to the focus of the reflector by a length of waveguide that is bent round so that it throws the energy at the face of the reflector, which is mounted on the satellite body so that it is always facing the Earth (Figure 35). The waveguide ends with a 'horn' which is matched to the size of the antenna, and which controls the shape of the cone of energy that it emits. In many cases, the horn is mounted on the centre-line of a circular antenna, supported by struts, known as a *centre-fed system* (Figure 36). These struts interfere with the antenna beam pattern, and centre-fed systems are being increasingly replaced by *offset systems*, in which the horn is kept out of the beam by being offset to one side of an elliptically-shaped antenna (Figure 37).

In theory, energy emitted from a single point at the focus of a parabolic antenna will result in a beam that is exactly parallel, as implied in Figure 35. In practice, this is not the case. The horn is not a single point, the parabolic shape is not perfect, and the parallelism of the beam is controlled by physical laws that are dependent on the frequency of the RF energy and the diameter of the antenna. At a typical frequency used for satellite transmissions, an antenna 1 m in diameter will produce a beam that is a cone of 2° angle. When this cone reaches the Earth 22 000 miles below, it will illuminate an area that is about 750 miles in diameter. The conic angle, and thus the diameter of the illumination, is exactly in inverse

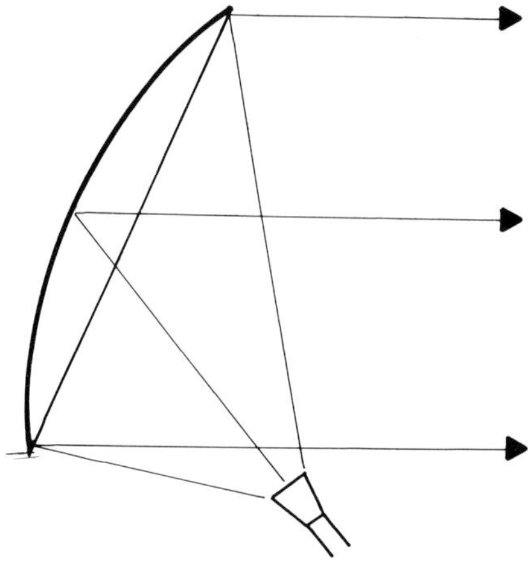

Figure 37 *An 'offset' antenna, where the antenna beam is not blocked by the feed horn*

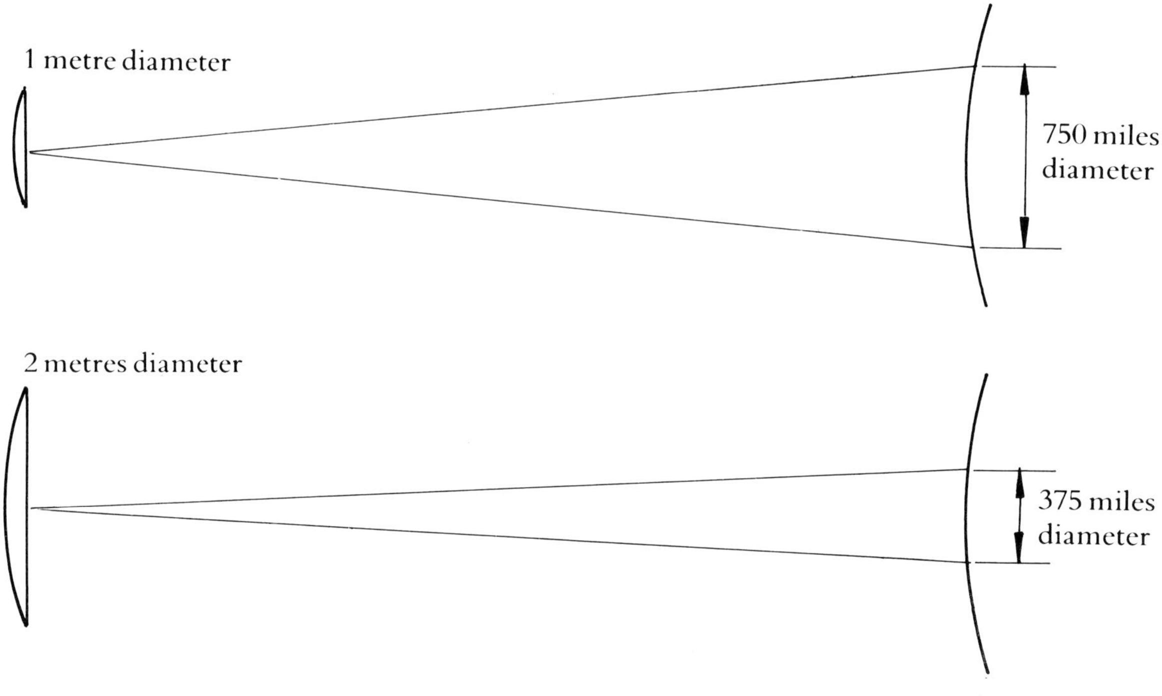

Figure 38 *At the same operating frequency, doubling the diameter of an antenna will halve the diameter of the beam at the Earth's surface*

proportion to the frequency and the diameter of the antenna. Thus, at the same frequency, as seen in Figure 38, an antenna of twice the diameter (2 m) will produce a 1° cone, and a 375-mile diameter illumination circle. An antenna of half the diameter (50 cm) will produce a 4° cone, and so on. Again, with the antenna fixed at 1 m diameter, doubling the frequency will produce a 1° cone, and halving the frequency will produce a 4° cone (Figure 39).

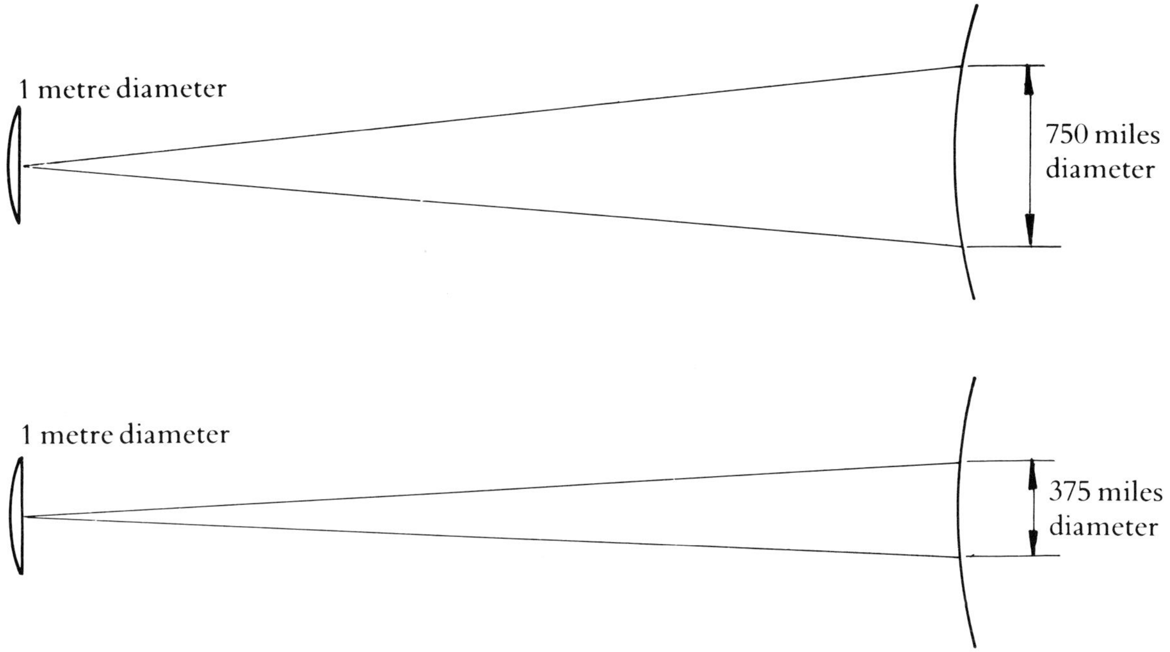

Figure 39 *With antennas of the same diameter, doubling the operating frequency will halve the diameter of the beam at the Earth's surface*

It can be seen, therefore, that in theory at least, coverage circles of any required size can be offered by using different frequencies and different antenna diameters. Again, in practice, this is not the case. The transmitted frequencies (or down-link frequencies as they are called) from satellites are strictly controlled by international agreement (as, indeed, are the up-link frequencies). If they were not, communications chaos would ensue. Thus, with the frequency specified, only the antenna diameter can be altered to change the coverage area on the ground. But, remembering the inverse-proportion law, it is clear that a very small coverage area – a small

Courtesy British Aerospace

Feed horns, clustered to provide a shaped beam, being tested in an anechoic chamber (one that absorbs stray radiation)

country, say – would need a very large antenna, perhaps several metres in diameter, and there are limitations on the size of antennas that can be packed into the confines of the launching rocket or the Space Shuttle. Antennas can, of course, be made to fold up inside the launcher and to open out to full diameter in space. Many folding and unfurling antennas are under development today, but they are very expensive in comparison with simple parabolic dishes, and their reliability in unfolding and locking into shape has yet to be proved.

Further, and unfortunately for the satellite designer, there are no circular countries or continents. Moreover, since satellites are often placed to the west of their service areas because of eclipsing, the conical beams from the satellites do not appear as circles on the Earth. They appear as distorted ellipses as the beam intersects the sphere of the Earth at varieties of angles (Figure 40). Thus, to achieve efficient coverage of a service area,

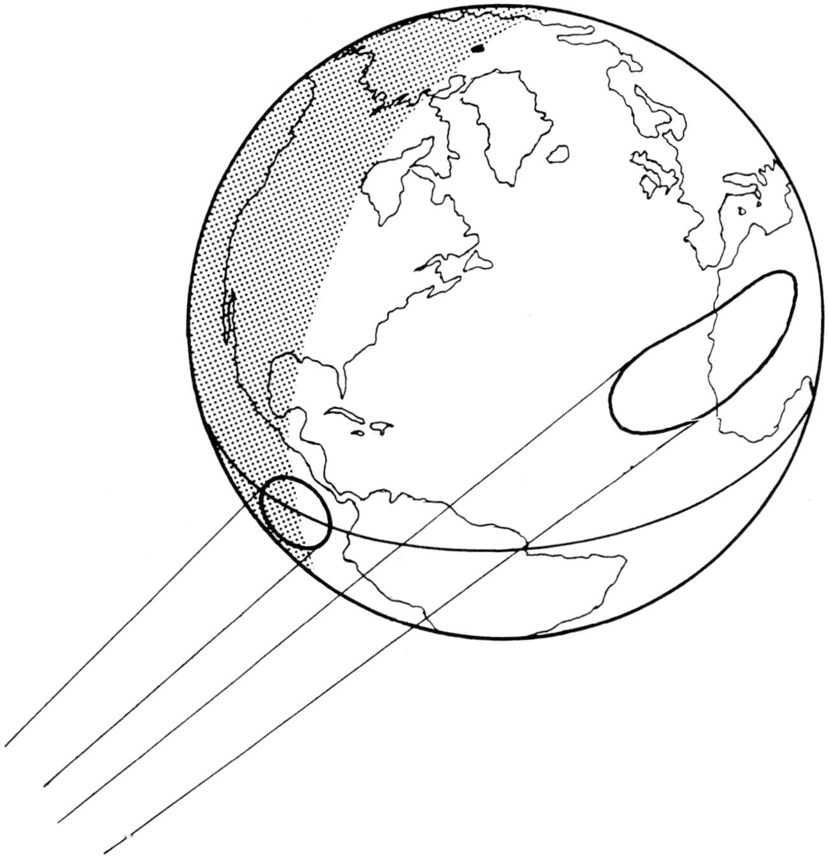

Figure 40 *A circular 'footprint', beamed down vertically from a satellite above the Equator, will become a distorted ellipse when beamed to another location (see coverage diagram in Figure 66, page 128)*

which may be a continent of complex shape such as Europe, or a 'long' country such as Italy or Chile, or an area in high latitudes, where the beam appears at a low angle from the geosynchronous orbit, the satellite designer must produce complex antenna shapes to match, often together with multiple horns to 'feed' the antenna from different angles.

Courtesy Matra SA

Earth view of the antennas on a Telecom 1 *satellite*

(Facing page)

Telecom 1 *satellite, with east and west cover panels removed, being set up for antenna tests in a radio-transparent dome on an outdoor test range*

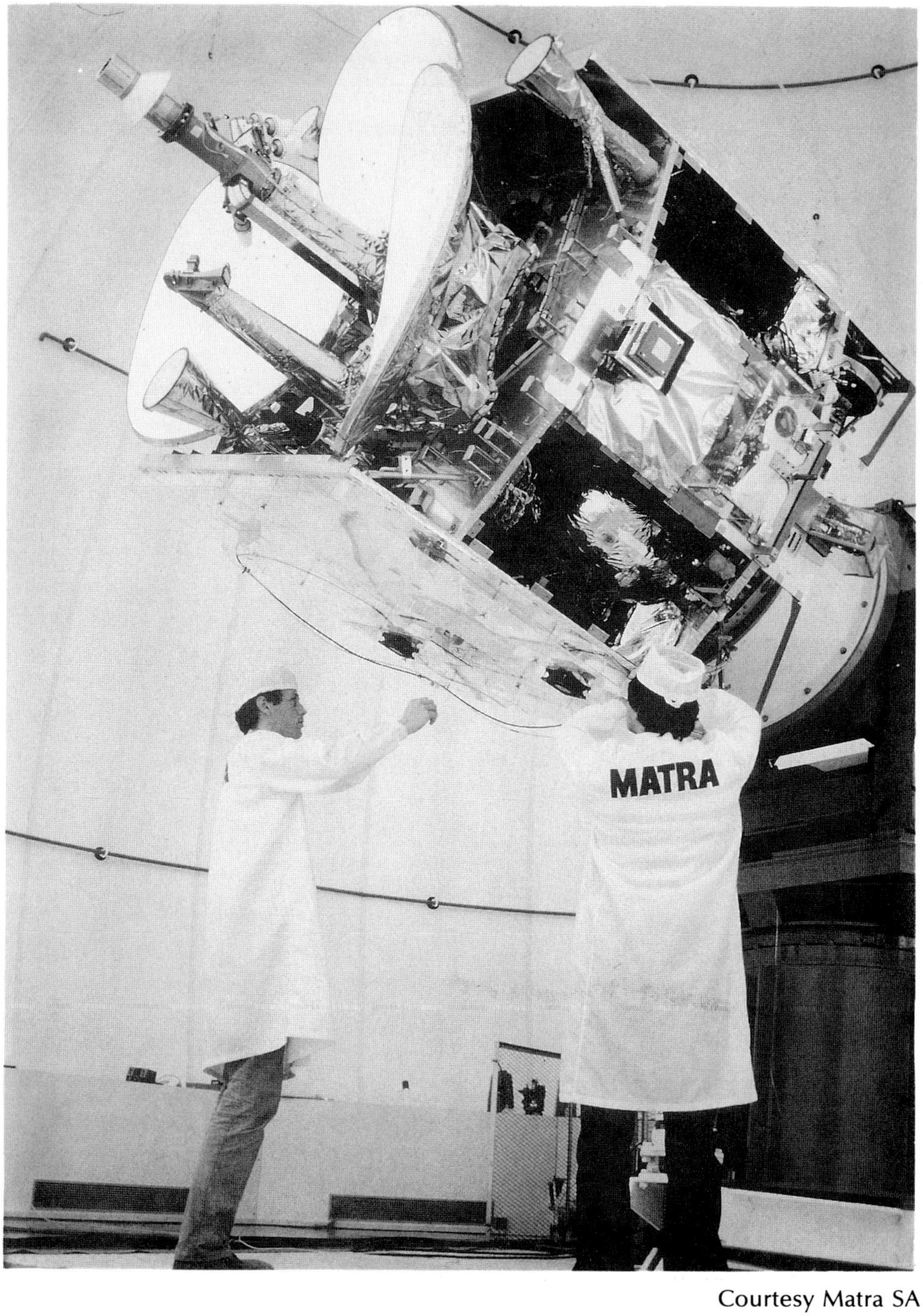

Courtesy Matra SA

Courtesy Matra SA

Telecom 1 *satellite in integration*

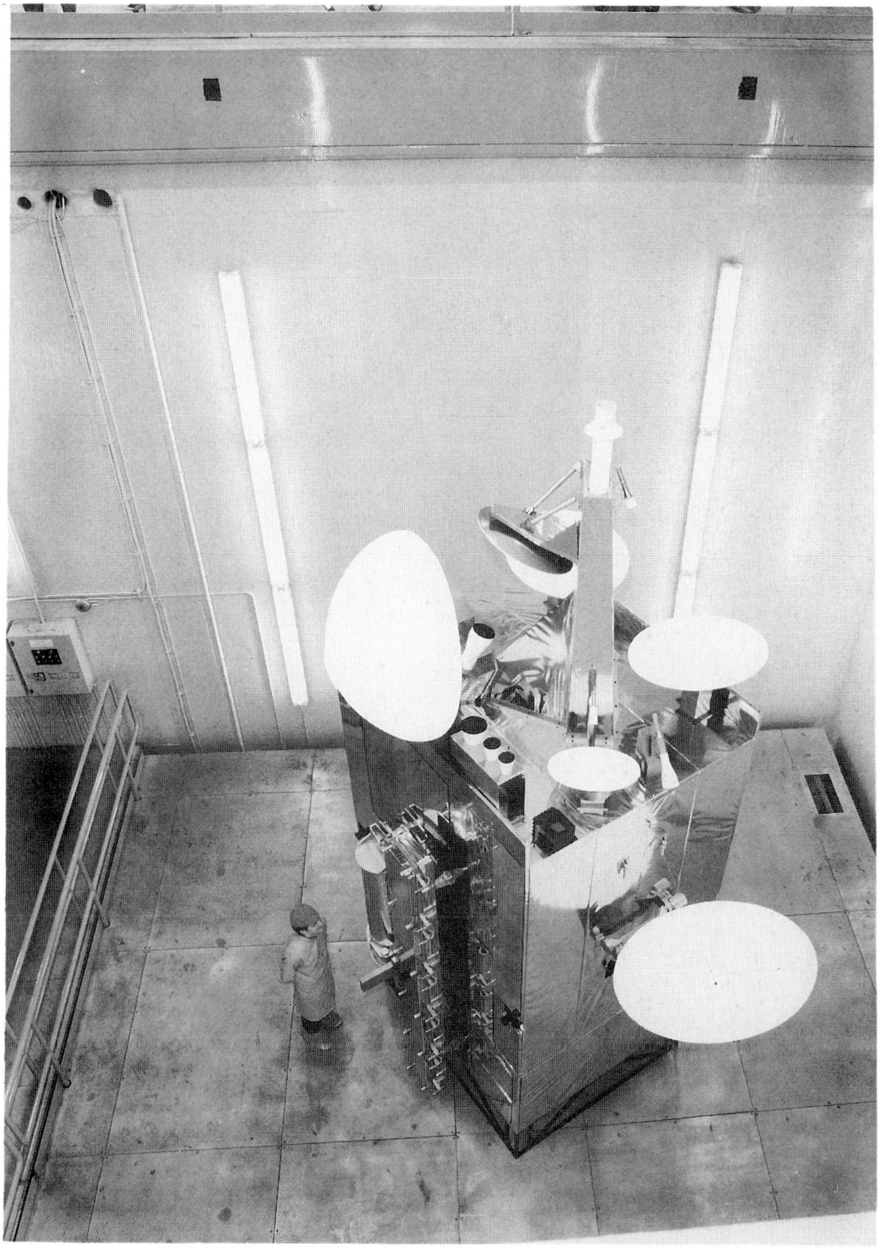

Courtesy British Aerospace

Prototype Olympus *satellite in the integration hall, showing the array of antennas for various services*

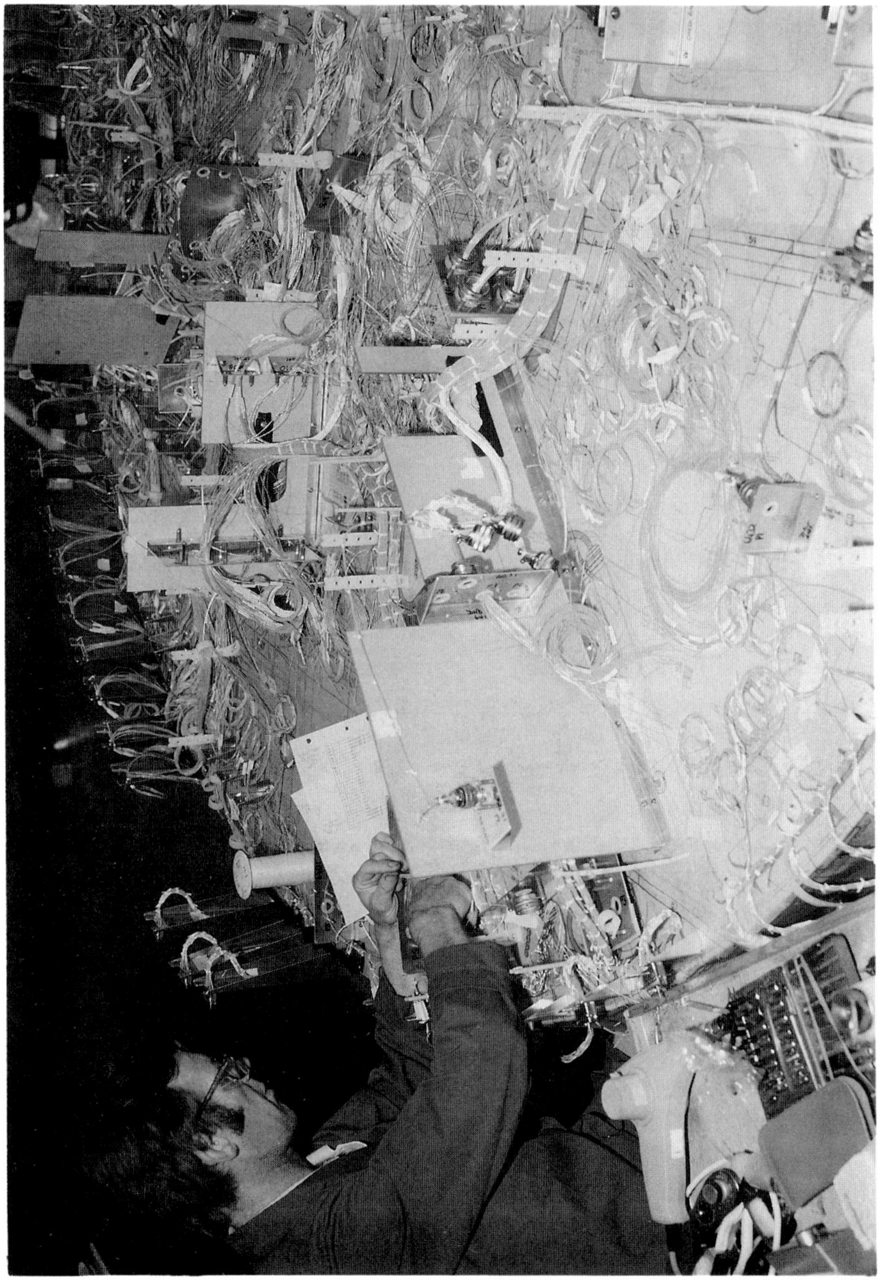

Courtesy British Aerospace

Part of the wiring assembly of the Intelsat VI *satellite*

Integration of an ECS-class satellite

Courtesy British Aerospace

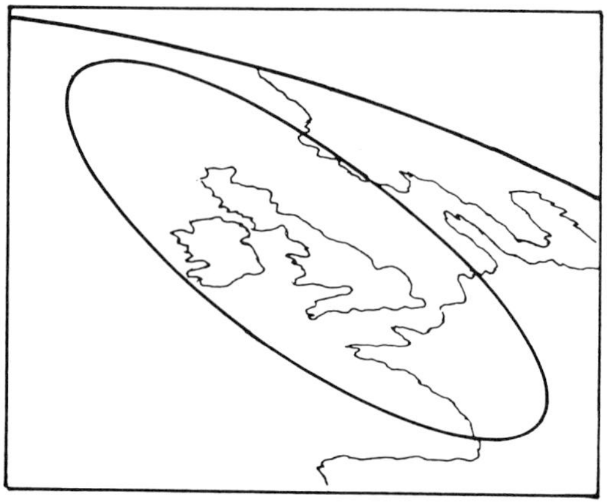

Figure 41 *A typical elliptical coverage requirement, seen from the satellite*

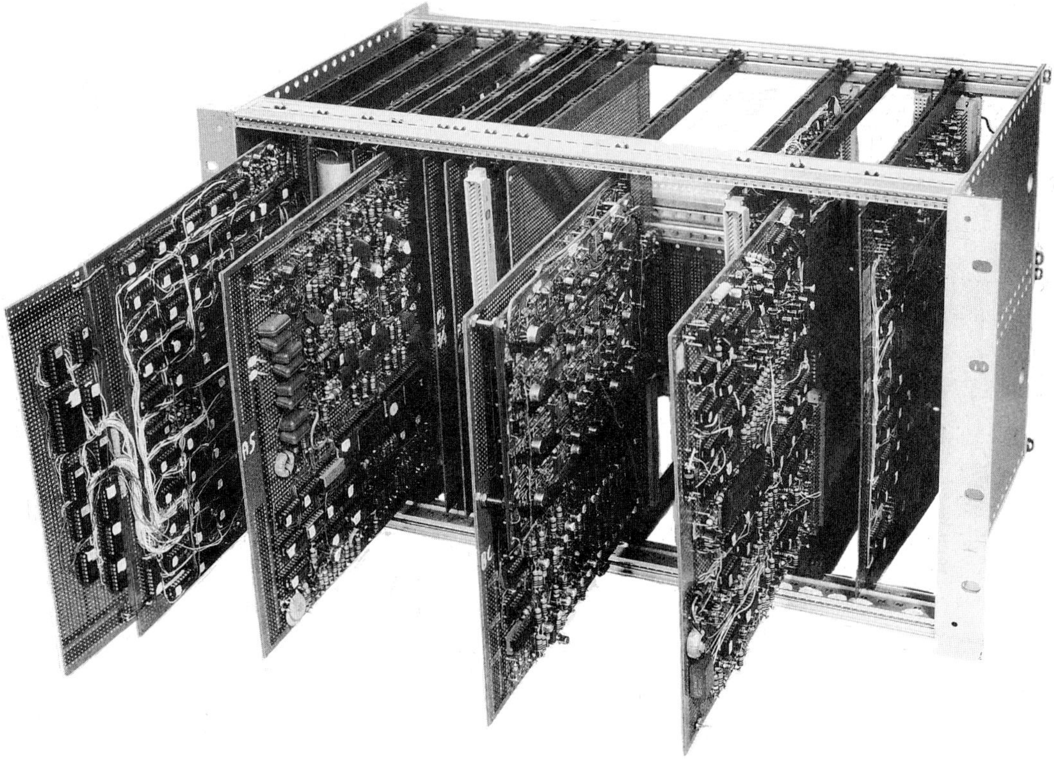

Courtesy British Aerospace

An on-board microprocessor for Olympus

Figure 41 shows, for example, an elliptical beam required to cover the British Isles from a satellite stationed in geosynchronous orbit above the Equator at longitude 31° West. This happens to be the allocated station, or *slot*, for satellites providing TV broadcasts for the United Kingdom. A centre-fed antenna to achieve this shape is seen in Figure 42, which shows that the antenna is an ellipse of exactly the same proportions as the beam, but rotated through 90°. Remembering the inverse-proportion law, it is seen that the long (major) axis of the satellite antenna will produce a narrow beam and the short (minor) axis will produce a wide beam, while all the other intermediate diameters of the antenna will produce beam diameters in inverse proportion. This results in an elliptical beam, shaped just like the antenna, but rotated through 90°.

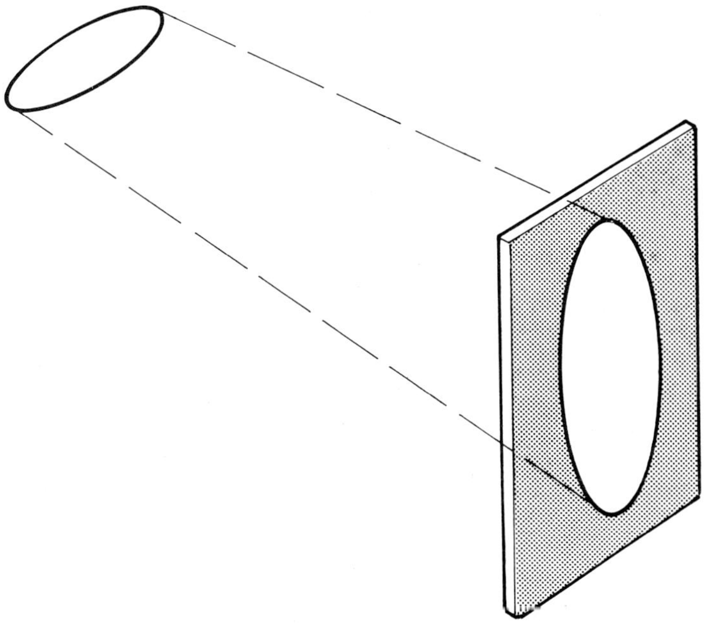

Figure 42 *An elliptical antenna produces an elliptical beam, rotated through 90°*

Another example is shown in Figure 43. Here, a triangular country needs to be covered. The 'best-fit' ellipse that can be generated by a centre-fed elliptical antenna will waste much of its energy on land or sea beyond the country's borders. Indeed, in some cases, such *spill-over*, as it is known, might give rise to serious political and economic repercussions from neighbouring countries. The spill-over is minimised by generating

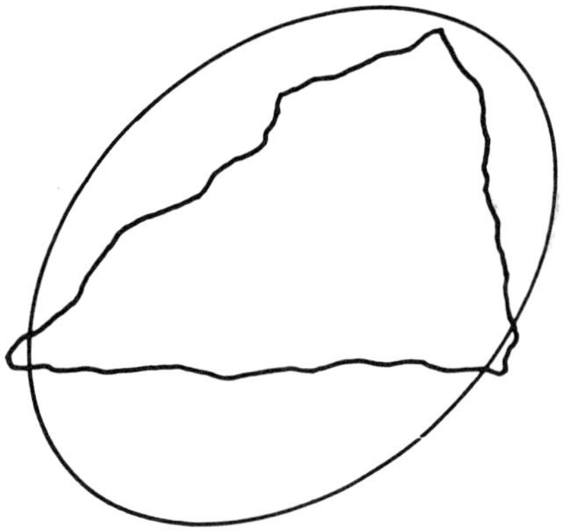

Figure 43 *Inefficient coverage of a country with a single elliptical beam*

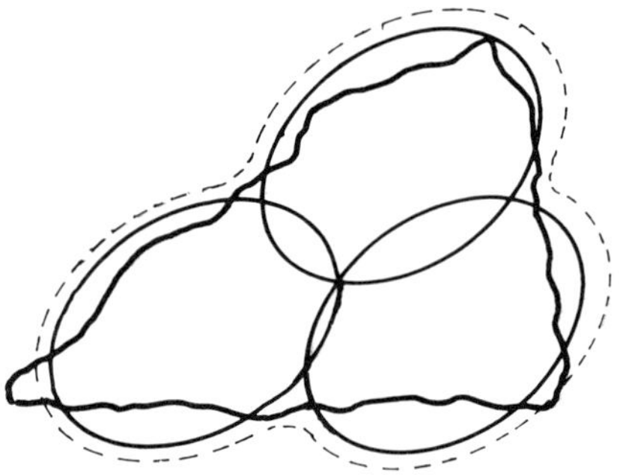

Figure 44 *Improved coverage efficiency with three beams produced by three feed horns reflecting off a single, larger antenna. The resulting total beam is indicated by the dotted line (see examples in Figures 68 to 78, pages 130–138)*

multiple beams (in this case three) and by feeding the antenna with three horns, each offset slightly from the antenna centre line, to give the overlapping pattern shown in Figure 44.

Spill-over also appears in another form. In all the references to beam sizes and beam shapes in this chapter so far, the impression is given that the edge of the beam has a sharp cut-off – that inside the beam there is full RF power, and that just outside it there is none. This is not the case. Earlier, it was seen that antenna feed horns are not theoretically perfect points of energy emission and that antennas themselves are not perfectly shaped. The result is that the final beam emitted by the antenna has an energy-distribution shape as shown in Figure 45. Full power appears on the centre line of the beam, falling to half-power at a small angle off centre and to zero power at the edge of the antenna. By convention, it is the half-power angle that is referred to as the beamwidth, and it is clear that more power appears outside that angle, falling more quickly as the angle is widened.

The effect of this spill-over, which is predictable mathematically, is shown in Figure 46 as an example. Here, the beam for the British Isles, first shown in Figure 41, includes the outer reaches of the beam as well as the half-power ellipse that covers the specified service area. Full power at beam centre falls on Lancashire in the north-west of England. Half power encompasses the whole of the British Isles including Ireland – this is the conventional 'edge-of-coverage' (EOC) line. Quarter power reaches Paris and the Low Countries, while one-tenth power covers most of France and

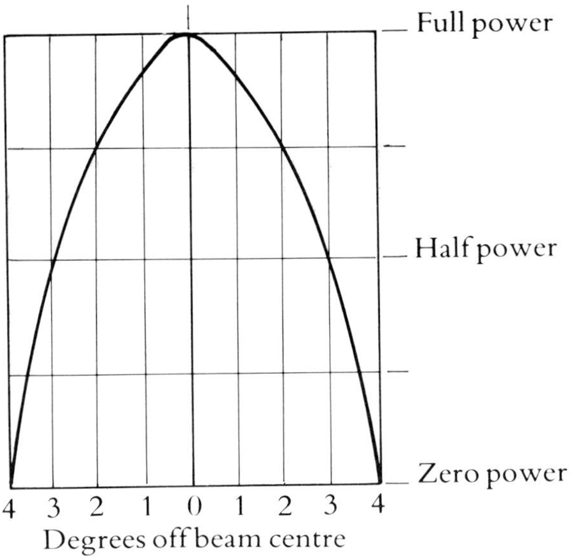

Figure 45 *Distribution of energy in an antenna beam*

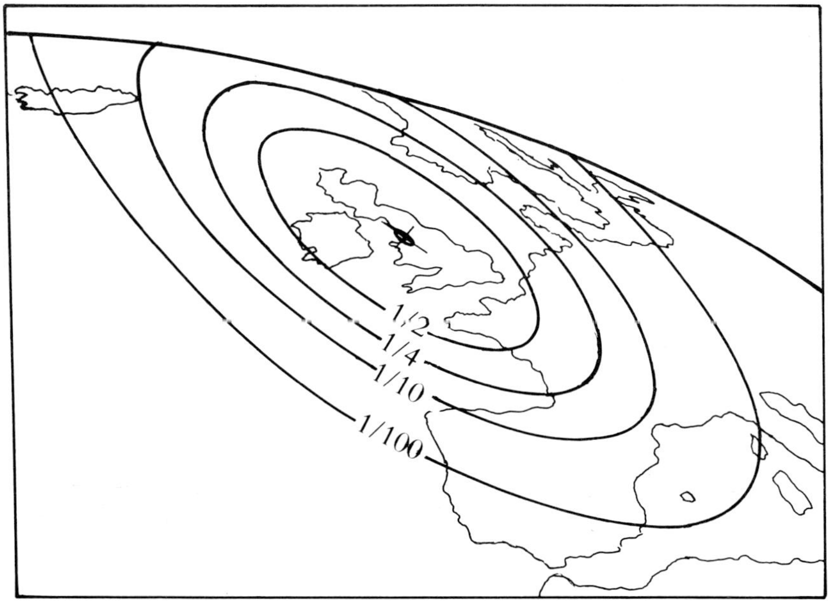

Figure 46 *The 'spill-over' effect of antenna energy distribution*

reaches into Scandinavia. After that, the beam power falls off rapidly until, in the Mediterranean and in the far north of Scandinavia, the power is down to one-hundredth of that at beam centre. Even at that low level, ground antennas of sufficient diameter – around 10 m – can capture enough power to provide an adequate TV picture. Ever-decreasing antenna sizes are required as the centre of the beam is approached until, at the centre, an antenna diameter of only 30 cm is needed.

These matters of power and antenna sizes, coupled with the frequencies used, are the essence of the characteristics of satellite communications, and are dealt with in the chapters that follow.

It is because of spill-over, and because it is necessary to guarantee a specified power level in the service area, that the satellite must remain pointing at the Earth with sufficient accuracy. A swing of the satellite body of 1° will move the beam centre and its EOC nearly 400 miles off target on the Earth's surface, with obvious repercussions. Normally, the pointing of the body is held to within a limit of less than one-tenth of a degree (40 miles on Earth). If even greater accuracy is required, the antennas can themselves be mounted on mechanisms that move them independently of the spacecraft body. While the satellite maintains its normal pointing accuracy, using its infra-red and other sensors as described earlier, the antennas lock on to radio signals emitted from Earth and maintain their more accurate pointing by using radar tracking techniques to drive their own pointing motors. These mechanisms can also be used to move antennas by several degrees to re-point them at different service areas.

Beam pointing and shaping can also be accomplished by fixed antennas that comprise a flat array of multiple emitters, each emitter being fed by an electronically-controlled distribution system. By changing the timing and phasing of the feeds to each emitter, the beam can be steered and changed in shape, and can even have 'holes' made within it. Phased array antennas of this type are used in military satellites where the holes are employed to blind jamming signals arriving from a particular area in the beam.

The complete satellite

Our crate has become a complete communications satellite. Now it can act as a communications and TV relay station covering areas as large as oceans and continents and as small as counties. But coverage area, energy delivered to the ground, the size of ground antennas, and the quantity of traffic that can be passed through the system are all closely interrelated. The choice of a particular satellite to handle a given traffic requirement is a matter of calculating those relationships and choosing the optimum solution. Before that optimisation process can be appreciated, it is necessary to understand the fundamental aspects of radio communications via space. These are seen in the next few chapters, which look at the simple basic mathematics that are needed, matters of radio frequencies, and the calculations that lead to the choice of a satellite.

7

Logarithms and decibels

Without the decibel (dB), the notebooks and computer screens of the satellite designer would be choked with thousands of zeros. Space communications uses numbers that are very large, in millions of millions, and very small – millionths or billionths. For instance, the loss of energy in a radio signal travelling from a satellite in geosynchronous orbit down to Earth is a ratio of 200 million million million, written in numbers as

$$\frac{1}{200\,000\,000\,000\,000\,000\,000}$$

Written in decibels, this is shown as −203 dB. Similarly, an amplifier might strengthen a radio signal by a factor of ten thousand million, that is, 10 000 000 000. In decibels, this appears as +100 dB.

Decibels not only act as a shorthand for large or small numbers, they ease the whole process of calculation because they replace multiplication and division of huge numbers by simple addition and subtraction of their much smaller equivalent numbers. They are a form of logarithms, which were invented by an English mathematician, Napier, for the very purpose of easing complex multiplication and division processes.

Logarithms can best be understood by looking at how they would be applied to the number 2. In the right-hand column of Table 1, the number of times that 2 is multiplied by itself (shown as 2^2, 2^3, 2^4 etc to the left) is called the *power of* 2, and that is the logarithm of the number in the preceding column. Thus, the logarithm (called 'log to the base 2' and shown as $\log_2$) of 16 is 4; the $\log_2$ of 512 is 9, and so on.

Examination of the table will show why logs are so convenient, even in this simple example. To multiply 32 by 128, the long multiplication process would give the correct answer of 4096. But the log of 32 is 5, and the log of 128 is 7. Simply add those together to give 12 and look for the number whose log is 12 – its *antilog*. It is 4096. Similarly, to divide 2048 by 64, take the log of 2048, which is 11, and deduct the log of 64, which is 6, to give a log of 5. The antilog of 5 is 32, which is the answer.

Table 1 *Logarithms to the base 2*

			Power of 2 or log_2
$2 \times 2 = 2^2$	=	4	2
$2 \times 2 \times 2 = 2^3$	=	8	3
$2 \times 2 \times 2 \times 2 = 2^4$	=	16	4
$2 \times 2 \times 2 \times 2 \times 2 = 2^5$	=	32	5
$2 \times 2 \times 2 \times 2 \times 2 \times 2 = 2^6$	=	64	6
$2 \times 2 \times 2 \times 2 \times 2 \times 2 \times 2 = 2^7$	=	128	7
$2 \times 2 \times 2 \times 2 \times 2 \times 2 \times 2 \times 2 = 2^8$	=	256	8
$2 \times 2 \times 2 \times 2 \times 2 \times 2 \times 2 \times 2 \times 2 = 2^9$	=	512	9
$2 \times 2 \times 2 \times 2 \times 2 \times 2 \times 2 \times 2 \times 2 \times 2 = 2^{10}$	=	1024	10
$2 \times 2 \times 2 \times 2 \times 2 \times 2 \times 2 \times 2 \times 2 \times 2 \times 2 = 2^{11}$	=	2048	11
$2 \times 2 \times 2 \times 2 \times 2 \times 2 \times 2 \times 2 \times 2 \times 2 \times 2 \times 2 = 2^{12}$	=	4096	12

Of course, there is no point in using logs for numbers as small as those in the table. They can be handled much faster by a hand calculator. But when it comes to billions and billionths, even a calculator cannot handle such numbers without recourse to logs. Moreover, when intermediate numbers have to be handled, say 3500 in the table above, for instance, the logs themselves are no longer simple whole numbers.

In handling daily calculations, it is easier and more practical to work to the base of 10 instead of the base of 2. Now we get

	Number	Log_{10}
$10^1 =$	10	1
$10^2 =$	100	2
$10^3 =$	1 000	3
$10^4 =$	10 000	4
$10^5 =$	100 000	5
$10^6 =$	1 000 000	6
⋮		

For intermediate numbers between 1 and 10, the log of 10 is less than 1, and each log is expressed as a decimal, as follows:

		Number	*Log₁₀*
10^{0}	=	1	0
$10^{0.3}$	=	2	0.3
$10^{0.48}$	=	3	0.48
$10^{0.6}$	=	4	0.6
$10^{0.7}$	=	5	0.7
$10^{0.78}$	=	6	0.78
$10^{0.84}$	=	7	0.84
$10^{0.9}$	=	8	0.9
$10^{0.95}$	=	9	0.95

Now, remembering that, to multiply numbers, the equivalent logs are simply added together, we can find the log of any number. For instance,

400 = 4 × 100
The log of 4 is 0.6
The log of 100 is 2
Therefore the log of 400 is 2.6

The log of 10 000 is 5
The log of 7 is 0.84
Therefore the log of 70 000 (= 7 × 10 000) is 5.84

In early audio practice, logs were referred to as *Bels*. In communications practice today, the unit used is one-tenth of a Bel, or one deciBel, written as dB. The dB is thus simply the log of a number multiplied by 10. So, taking the two examples just given: the log of 400 is 2.6; this is equal to 26 dB. The log of 70 000 is 5.84; this is equal to 58.4 dB.

Values in dB will be used throughout the rest of this book. Tables 2 and

Table 2

Numbers in tens	dB
10	10
100	20
1 000	30
10 000	40
100 000	50
1000 000	60

Table 3

Unit numbers	dB
1	0
2	3
3	4.8
4	6
5	7
6	7.8
7	8.4
8	9
9	9.5

3 give decibel equivalents of multiples of ten and of units. It will be seen that for multiples of ten the dB value is simply 10 × the number of zeros after each '1'. So 100 dB is 10 000 000 000.

We can now take new examples, working in dB:

The number 10 000 is 40 dB
The number 6 is 7.8 dB
Therefore 60 000 is 47.8 dB

37 dB = 30 dB + 7 dB
30 dB is 1000
7 dB is 5
Therefore 37 dB is 5 × 1000 = 5000

99 dB = 90 dB + 9 dB
90 dB is 1000 000 000 (nine zeros)
9 dB is 8
Therefore 99 dB is 8000 000 000

Of course, there are other intermediate numbers that are not shown in Tables 2 and 3. What, for instance, is 32 dB? 2 dB is not shown in the tables. Figure 47 shows the relationship of numbers to dB for all numbers from 1 to 10. It will be seen that 2 dB is equivalent to 1.6, so 32 dB is 1000 × 1.6 = 1600.

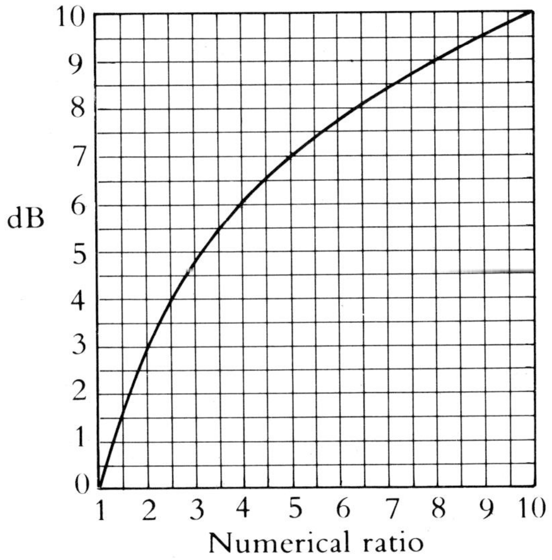

Figure 47 *Relationship between dB and numerical ratio, from 0 to 10 dB*

Accurate dB equivalents from 1 to 10 can be taken from log tables or, more easily, from hand calculators that offer a log facility. Using a calculator, key the number, key 'log' and then multiply the answer by 10. Working the other way, that is to obtain the accurate value of a given dB, divide the dB by 10, key 'inverse' and then key 'log'.

In all the examples shown so far, the numbers have been greater than 1, and the equivalent dB values have been positive. For numbers less than 1, the equivalent negative dB value is used. For example:

100 is 20 dB

$\frac{1}{100}$ (one hundredth) is -20 dB

5000 is 37 dB

$\frac{1}{5000}$ is -37 dB

Positive and negative dB can be simply added arithmetically to give the final answer. Thus

$$80\,000 \times \frac{1}{400} \quad \text{is} \quad 49 \text{ dB} - 26 \text{ dB} = 23 \text{ dB} \qquad \text{which is } 200$$

That is a simple sum. It is when the communications designer finds that he has to divide 24.8 thousand billion by 87.6 million that he thanks Napier and his successors for dB.

In the daily jargon of space communications, dB are used as freely as their equivalent numbers, simply because the designers work in dB. Thus one hears, 'You'll lose 3 dB that way'. Losing 3 dB means -3 dB, which means 1/2. The equivalent phrase would thus be 'You'll lose half the power that way.' Similarly, 'That is 7 dB up' means that it is up by a factor of 5.

Usually, in communications, the dB is applied to power in watts, and the suffix 'W' is placed after the dB to indicate this usage. Thus, 20 dBW means 100 watts, and 23 dBW means 200 watts. Similarly, -100 dBW means one-tenth of a thousandth of a millionth of a watt.

The dB is used also as a ratio in some calculations. For instance, the gain of an antenna is always referred to in relation to isotropic radiation and is thus shown in dBi.

To those whose daily work involves the continuous use of dB, it can be used for a variety of other purposes as well. There has been seen, on a note passed from a communications engineer to a girl in his laboratory, the legend '30 dBx'. She read it correctly as 'a thousand kisses'.

8

Radio frequencies and wavelengths, and their allocations

The terms used to express frequency and wavelength are

hertz (Hz) for frequency, in cycles per second
λ (the Greek letter lambda) for wavelength, in metres (m)

The relationship between Hz and λ is given by

$$\lambda = \frac{3 \times 10^8}{\text{Hz}}$$

Thus, for example, if the frequency is one million cycles per second (10^6 Hz), $\lambda = 3 \times 10^2$ m, or 300 metres. At a frequency of a billion cycles per second (10^9 Hz), $\lambda = 0.3$ metres.

The unit for frequency carries a prefix to designate ranges of 1000 Hz:

1000 Hz is 1 kilohertz, shown as 1 kHz	$\lambda = 300\,000$ metres
1000 000 Hz is 1 megahertz, shown as 1 MHz	$\lambda = 300$ metres
1000 000 000 Hz is 1 gigahertz, shown as 1 GHz	$\lambda = 0.3$ metres

The 'long waves' in the kHz range, used in terrestrial communications, are not suitable for communications via space, nor are the 'medium' and 'short' waves in the low MHz bands. They need antennas that are far too large for space use, and they are subject to fading and distortion in the Earth's ionospheric and magnetic fields. The lowest frequency used in satellite communications is around 800 MHz, and the highest today is around 30 GHz. Within this range of 'microwave' frequencies, the bands that are the most widely employed are usually referred to by identification letters (Table 4).

By international agreement, under the auspices of the International Telecommunication Union (ITU) in Geneva, the uses of these bands are allocated to various purposes in different parts of the world. First, the world is divided into three regions, as shown in Figure 48. Then, within these regions, the uses are divided into three different types of service.

Table 4 *Band names used in satellite communications*

Frequency range	*Band name*
1 – 2 GHz	L-band
2 – 3 GHz	S-band
4 – 6 GHz	C-band
7 – 8 GHz	X-band
11 – 18 GHz	Ku-band (*kay-you*)
20 – 30 GHz	Ka-band (*kay-ay*)

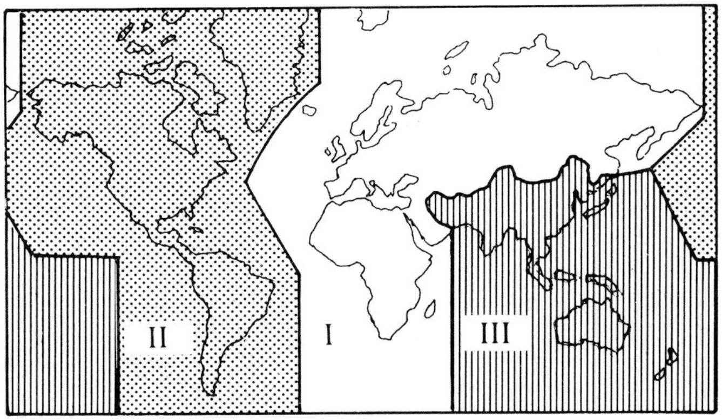

Figure 48 *World telecommunications regions, designated by the International Telecommunication Union (ITU), Geneva*

Fixed services

Fixed services comprise voice, data and video communications between Earth stations that are sited at fixed locations. In Europe (Region 1) such stations are generally owned by the national Post, Telecommunications and Telegraph authority (PTT) of the country concerned, and all such stations must be registered with and approved by the appropriate authority. Only the United Kingdom permits a competitive operator (Mercury) to own fixed service stations for both transmission and reception. In the United States (Region 2) fixed service stations for transmission and reception may be owned by any operator, subject to the approval of the Federal Communications Commission (FCC).

Transportable stations, which can be moved from site to site, are considered as fixed service stations when they are in operation at the temporary locations.

Receive-only stations, used by cable-TV operators to receive programmes from a fixed service satellite for redistribution to homes on a cable network, are also considered as fixed service stations and must be appropriately registered.

Broadcast services

Broadcast services comprise voice, data and video communications from one or more fixed stations to any number of receive-only stations located anywhere in the coverage area of the satellite. Ownership of the fixed transmitting stations may be public or corporate and must be registered. The receive-only stations may be owned by anyone – corporate or private – and do not need to be registered or approved.

This class of service is aimed primarily at the broadcasting of high-power TV signals directly into small antennas in the home. It also includes high-quality audio broadcasting, and the sending of commercial data to 'multi-drops' – for example, the broadcasting of latest stock and price data to a corporate sales and service network that covers the whole of Europe.

The service can, of course, be received by cable-TV operators, who can then redistribute the programmes to their cabled customers. In some cases, regulations call for 'must carry', that is, the operators must offer the broadcast service together with their other programmes. In other cases, cable operators must not redistribute the service without authority to do so.

Mobile services

Mobile services comprise voice and data communications between fixed stations and mobile users in ships, trucks and aircraft, and communications between mobile users via fixed stations. Higher satellite powers of the future will permit personal, hand-held transceivers to be embodied in these services.

The frequency bands are then allocated to the services as shown in Table 5, using different frequencies for up-links (Earth to satellite) and down-links (satellite to Earth). Table 5 applies to Region 1, Europe.

Today, the total available bandwidth of 1100 MHz in C-band is taken up by international communications through the Intelsat series of satellites. In the United States, most domestic satellite links operate in C-band as well, and the allocated bandwidth is also filled. Accordingly, space communications links are being moved into Ku-band, which has been in operation in Europe since the late 1970s, and which is now emerging in the United States and for international communications. Ku-band brings the advantage of higher gain in an antenna (see Chapter 6) which permits the use of smaller antennas to handle a required traffic rate, but the band suffers more than C-band from the attenuation effects of the

Table 5 *Allocation of radio frequencies to satellite communications in Region 1 – Europe*

Band	*Approximate frequency range* (GHz)	*Link*	*Service*	*Available bandwidth* (MHz)
L	1.5 – 1.6	Down	Mobile	100
	1.6 – 1.7	Up	Mobile	100
S	2.5 – 2.6	Down	Broadcast	100
C	3.4 – 4.2	Down	Fixed	800
	4.5 – 4.8	Down	Fixed	300
	5.9 – 7.0	Up	Fixed	1100
X	7.2 – 7.7	Down	Military	500
	7.9 – 8.4	Up	Military	500
Ku	10.7 – 11.7	Down	Fixed	1000
	11.7 – 12.5	Down	Broadcast	800
	12.5 – 12.75	Down	Fixed*	250
	12.75 – 13.25	Up	Fixed*	250
	14.0 – 14.8	Up	Fixed	800
	17.3 – 18.3	Up	Fixed	1000
Ka	17.7 – 20.2	Down	Fixed	2500
	20.2 – 21.2	Down	Mobile	1000
	22.5 – 23.0	Down	Broadcast	500
	27.0 – 30.0	Up	Fixed	3000
	30.0 – 31.0	Up	Mobile	1000

*Allocated to 'business services' through antennas located at customer premises.

Notes:
Available bandwidth in megahertz is the frequency range in gigahertz mutliplied by 1000.
In each band, total available bandwidth on the down-link is equal to that on the up-link.

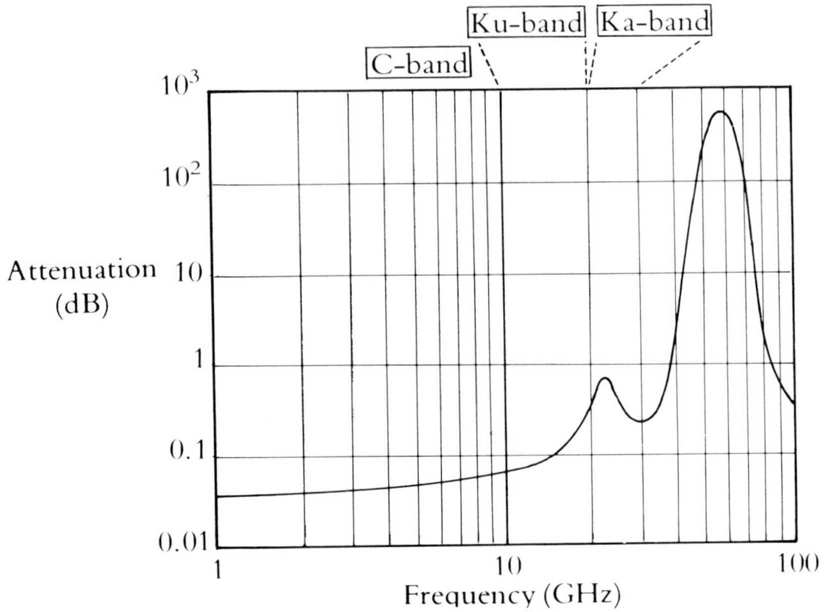

Figure 49 *Atmospheric attenuation plotted against operating frequency*

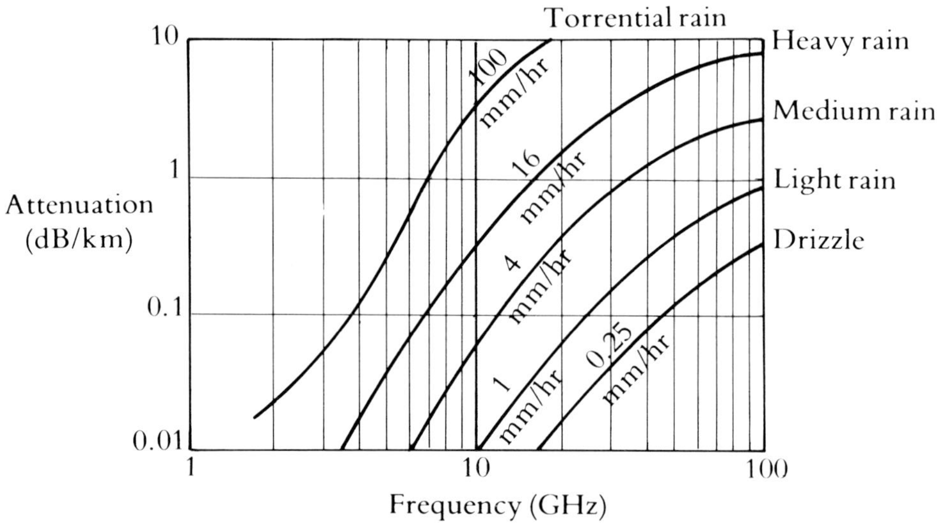

Figure 50 *Rain attenuation plotted against operating frequency*

atmosphere and rainfall (Figures 49 and 50). However, these losses are more than compensated by the ability of a Ku-band Earth station to handle more traffic than a C-band station with the same size of antenna, and the band offers more available bandwidth than C.

Both C-band and Ku-band suffer from problems of interference with terrestrial microwave systems, which share the same frequency bands, though not usually exactly the same frequencies. This is the reason for the limitations in bandwidth made available to space communications. These problems do not apply to Ka-band, which is now entering the space scene. The table shows that this band offers a much wider available bandwidth (4000 MHz against the 2050 MHz of Ku and the 1100 MHz of C), which means that it can handle proportionately more traffic. However, as Figures 49 and 50 show, this band suffers more than the others from atmospheric and rain attenuation, and allowance for these losses must be made in calculating traffic capacities. Nevertheless, as overall traffic demands increase, a move into Ka-band will be inevitable.

9

Coverage, gain and EIRP

The purpose of satellite antennas was explained in Chapter 6. They are provided to capture the radio signals emitted by the satellite and to reflect them on to areas on Earth where they are required. These are the coverage areas. The action of reflecting the signals and forming them into a beam intensifies them, and the amount of intensification is called the 'gain' of the antenna.

EIRP is a product of the RF power of the satellite and its antenna gain. EIRP stands for *equivalent isotropic radiated power*; sometimes, *effective* is used instead of *equivalent*. It is a term used universally in satellite communications because it is the one component that provides a measure of the quality of the down-link service offered by any satellite, independently of its coverage and its power. It thus enables suppliers and users to compare the qualities of competing satellite systems without having to worry about what areas they cover or what power the satellites produce.

EIRP can best be understood by comparing three different satellites (Figure 51). Satellite A carries a large antenna that forms a narrow beam. When the beam reaches the Earth, it 'illuminates' an area of, say, 100 000 square kilometres (that is, a circle of about 350 km diameter). Each transponder that is supplying RF power to the beam produces 20 watts.

Satellite B carries an antenna of which the diameter is half that of the one on Satellite A. Because, at the same frequency, the conical beam produced by an antenna is inversely proportional to the antenna diameter, Satellite B will produce a beam that is twice the size of A, that is, about 700 km diameter; this is an area of about 400 000 square kilometres – four times that of A.

Satellite C carries no antenna at all. All its RF power is radiated out into space in spherical waves of increasing size. When a wave front reaches the

Earth 22 000 miles below, only a tiny fraction of it will be captured by the Earth, which will be illuminated over its entire face – an area of many million square kilometres.

With Satellite A, the 20 watts of RF power will be spread over 100 000 square kilometres, and this will result in a certain flux density, measured in watts per square metre, which will determine the quality of the service. To provide an *equivalent* service, Satellite B must produce four times the power, i.e. 80 watts, because its power is being spread over four times the area. Again to provide an *equivalent* service, Satellite C must produce

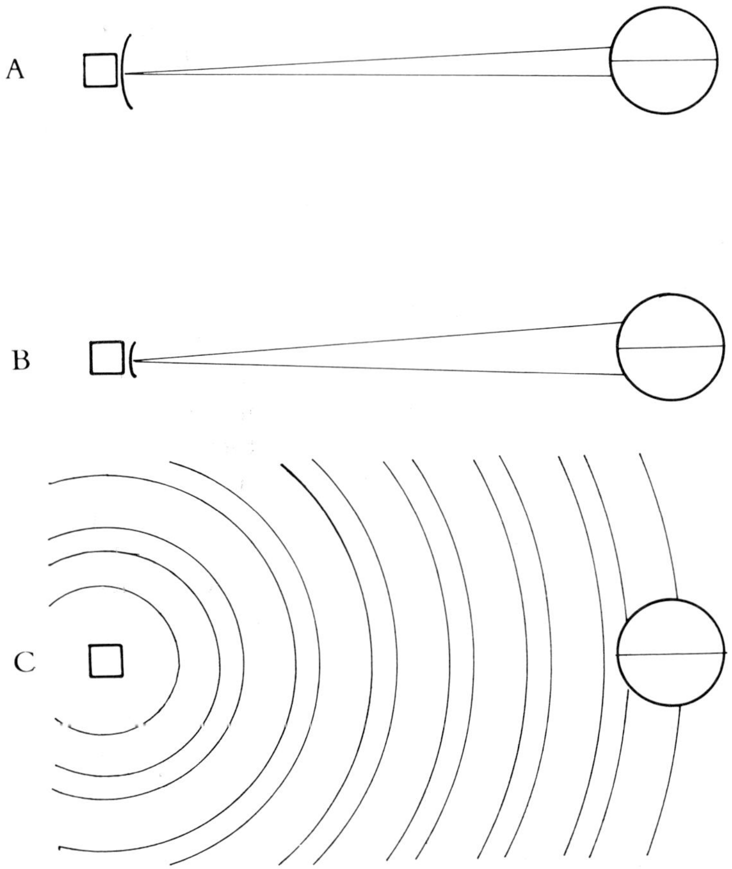

Figure 51 *Equivalent isotropic radiated power (EIRP). Satellite* A *carries an antenna that provides a certain coverage diameter on the Earth. Satellite* B *carries an antenna half that of* A, *and provides a coverage area of twice the diameter. Satellite* C *carries no antenna – its power is radiated 'isotropically', and only a very small fraction of it reaches the Earth. It is Satellite* C *that is used as the reference for the EIRP of the other two satellites*

power measured in thousands of watts. But this is the satellite that carries no antenna – it is radiating *isotropically*. It is independent of coverage area.

Satellite C becomes the standard against which the other two satellites are measured, because the user on Earth simply wants to know what flux density he will receive for his service, regardless of whether its source is a weak satellite with a large antenna, a stronger satellite with a small antenna, or a still more powerful satellite with no antenna at all. For making comparisons, and to calculate the quality of his service and the amount of traffic it can handle, he simply uses the EIRP of the service being offered.

If, for instance, a cable TV operator owns a 3 m diameter ground antenna and finds that he gets good quality signals from a satellite offering an EIRP of 40 dBW, he will know that another satellite offering 43 dBW will allow him to use an antenna of half the area, that is, just over 2 m in diameter. Similarly, the 43 dBW satellite will offer twice the power of the 40 dBW satellite to an antenna used for telecommunications, enabling it to handle twice the traffic of the same quality.

Note that 40 dBW means 10 000 watts of RF power, and 43 dBW means 20 000 RF watts. As shown in Chapter 5, this means that the two satellites would have to generate 30 kW and 60 kW respectively from their solar arrays if they were isotropic radiators without antennas. This is ten to twenty times more than the power produced by the largest satellites in use today and is clearly impractical. But these powers are not real – they are the *equivalent isotropic* powers. In reality, the satellites' antennas will provide gain that will bring the real power down to practical levels.

The gain of an antenna is directly proportional to its diameter, and inversely proportional to the wavelength of the frequency being used. The working formula for gain is

$$G = e\left(\frac{\pi D}{\lambda}\right)^2$$

where e is the efficiency of the design, usually taken as 55%
π is the constant 3.14
D is the antenna diameter in metres
λ is the wavelength in metres.

Putting in the values for e and π, this reduces to

$$G = 5.4\left(\frac{D}{\lambda}\right)^2$$

Take an example of an antenna 1 m in diameter, working at 12 GHz:

$$D = 1 \qquad \lambda = 0.025$$

Then

$$G = 5.4 \times 40^2 = 8640 = 39.4 \text{ dBi}.$$

As another example, take a 1.5 m antenna working at 4 GHz:

$$D = 1.5 \qquad \lambda = 0.075$$

Then

$$G = 5.4 \times 20^2 = 2160 = 33.3 \text{ dBi}.$$

Figure 52 shows a range of gains against varying diameters and frequencies. To obtain EIRP, all that is necessary is to add these gains to the output power of the satellite amplifiers, also expressed in dBW. If, in both examples, the output power per transponder is 20 watts, that is 13 dBW, the EIRP of the first example is 39.4 + 13 = 52.4 dBW. The EIRP of the second example is 33.3 + 13 = 46.3 dBW.

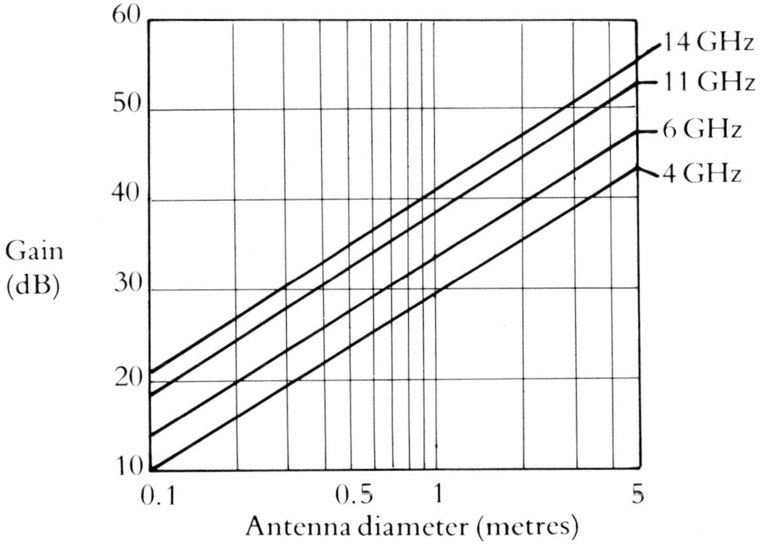

Figure 52 *Antenna gain plotted against diameter and operating frequency*

Note that, in the first example, the EIRP of 52.4 dBW could have been obtained with a smaller antenna operating at higher power. If, for instance, the antenna were only 0.5 m in diameter, its gain at 12 GHz would be

$$5.4 \times 20^2 = 2160 = 33.3 \text{ dBi}.$$

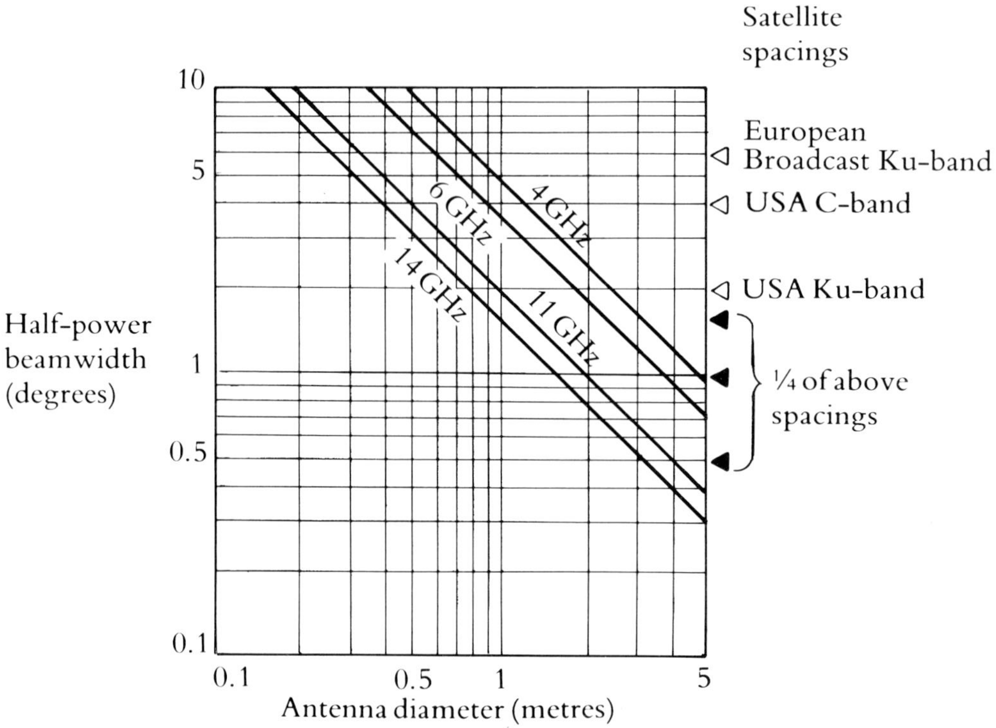

Figure 53 *Antenna half-power beamwidth plotted against diameter and operating frequency. The quarter-spacing marks indicate the minimum antenna diameter that should be used for the services shown*

Then, to obtain the same EIRP of 52.4 dBW, the output power would have to be

$$52.4 - 33.3 = 19.1 \text{ dBW} \qquad \text{which is 80 watts.}$$

This is clearly the case because, in halving the diameter of the antenna, its gain has been reduced by a factor of $4(D^2)$ so, to maintain the same EIRP, the power must be increased by a factor of four.

It is not practical, however, to increase or decrease EIRP simply by changing antenna diameters and output powers at will. In Chapter 6 it was shown that the conical beam produced by a circular antenna is dependent on the antenna diameter, and the size of that beam determines the coverage area that will be seen on Earth. Thus it is the coverage area that fixes the size of the antenna and, once that is fixed, EIRP can be altered only by changing output power.

The relationship between the size of the conical beam and antenna diameter is given by the formula:

$$\theta = 70\frac{\lambda}{D}$$

where θ is the half-power beamwidth in degrees
λ is the operating wavelength in metres
D is the antenna diameter in metres.

Thus, the beamwidth of a 1 m diameter antenna operating at 12 GHz (λ = 0.025 m) is

$$\theta = \frac{70 \times 0.025}{1} = 1.75°.$$

Figure 53 shows a range of beamwidths plotted against varying diameters and frequencies.

In the simplest case of a satellite stationed in geosynchronous orbit vertically above its service area (at the same longitude on the Equator) a 1.75° beam emitted by the satellite would produce a circular 'footprint' on the Earth of about 670 miles (1000 km) diameter, that is, a land area about the size of the British Isles. If the satellite had to serve an area twice that diameter (four times the area) at the same frequency, the antenna diameter would have to be reduced to 50 cm diameter. Its gain would fall by a factor of four (exactly the inverse of the increase in coverage area) and with the same output power the EIRP would fall by a factor of four as well. To offer the same EIRP over the larger area, the output power would have to be increased by a factor of four.

It can be seen, then, that coverage area is the first determining factor in selecting a satellite system, for it fixes the size of the antenna and thus its gain. Once gain is fixed, EIRP follows from the power produced by the satellite, and it is EIRP that will determine the performance of the whole communications network. The network itself will be using Earth stations of varying quality and capacity, and they must be considered too before the final satellite system is adopted. The next chapter looks at Earth stations and their impact on quantity and quality of communications traffic.

10

Earth stations

Types, construction and sizes

There are two types of Earth station – those that receive only and those that can both receive and transmit. Receive-only stations are used principally for the reception of television signals emitted by satellites, when they are usually known as TVRO stations. They are also used for receiving data and other forms of information that can be displayed visually or in printed form. For two-way links between users, such as telephony, video-conferencing and computer tie-ups, the stations at each end of the links are provided with both transmission and reception facilities.

Both types of station employ similar geometries to capture and focus the signals arriving from the satellite (the *down-link* signals) and to aim signals at the satellite (the *up-link*). Figure 54 shows how the down-link signals are captured by the Earth station antenna, which usually takes the

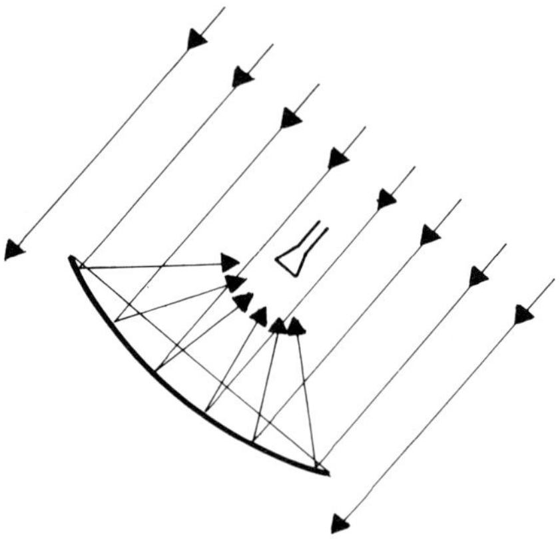

Figure 54 *Geometry of a centre-fed Earth station antenna*

Courtesy Matra SA

A 90 cm (3 ft) diameter offset Earth station for data transmission and reception

form of a parabolic circular dish and which reflects the captured signals to the focus of the parabola, where a collecting 'horn' is mounted. For transmission purposes, the horn emits signals which are reflected off the parabola to form the up-link beam.

Mounting a horn in this manner, at the focus of a circular dish, means that a structure has to be provided for the mounting, and this structure interferes with the clear line-of-sight of the dish, reducing its efficiency. For this reason, many new stations now employ parabolic dishes that are elliptical in frontal view with the horn offset from the centre-line of the dish (Figure 55). The dishes are elliptical so that they present an apparent circular shape as 'seen' by the satellite, to give a circular cone into the horn.

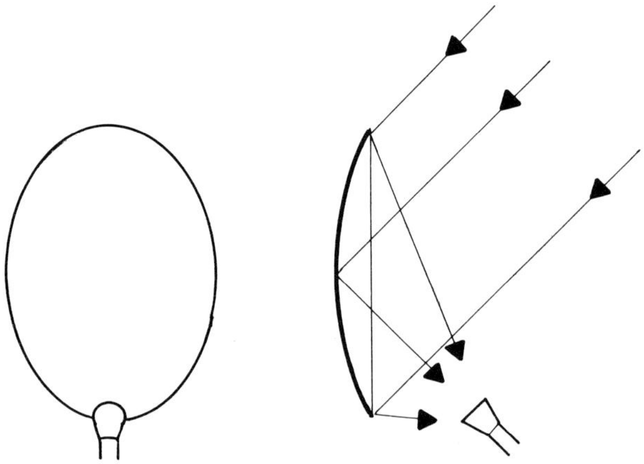

Figure 55 *Geometry of an offset Earth station antenna. It is elliptical in frontal view to present a circular shape to the satellite*

Signals entering the horn – or leaving it – must be conducted to and from the amplifying equipment in the station. This conduction takes place through a length of waveguide – rectangular-section tubing of a size dependent on the operating frequencies. In some designs of small stations this waveguide is used as the horn-mounting structure to minimise interference (Figure 56). To minimise losses in conduction, the waveguide length must be kept as short as possible. This is easier in offset designs than with centre-fed circular dishes. The shortest length of all is achieved with what is known as a *Cassegrain antenna* (Figure 57). Here, by using double reflectors, the amplifying equipment can be mounted directly behind the horn. This type of antenna is usually seen, however, in only

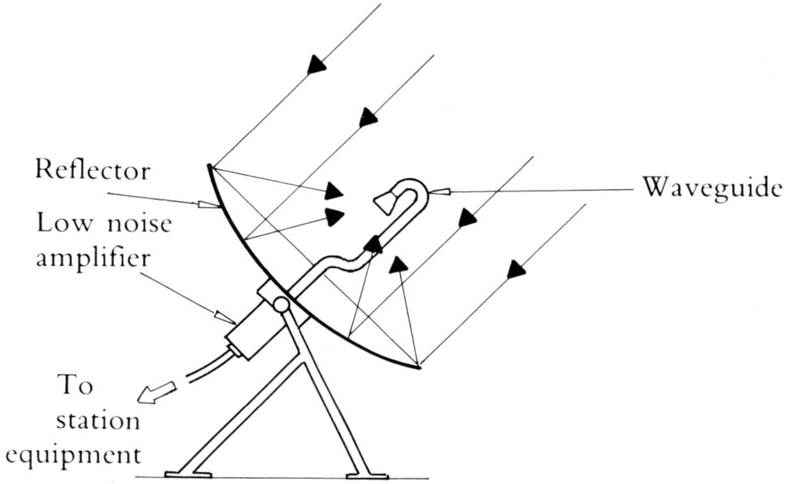

Figure 56 *A typical small Earth station, in the range 1 to 2 metres diameter, with the feed horn mounted on its own waveguide*

the largest and most expensive Earth stations where maximum performance is essential.

Figure 58 shows the principal components of the transmission and reception equipment in a typical small Earth station used for voice, data and video exchange in a Ku-band business service. On the transmit side, incoming digital signals are mixed together in a multiplexer to produce a combined data stream which is then passed into a modulator. The

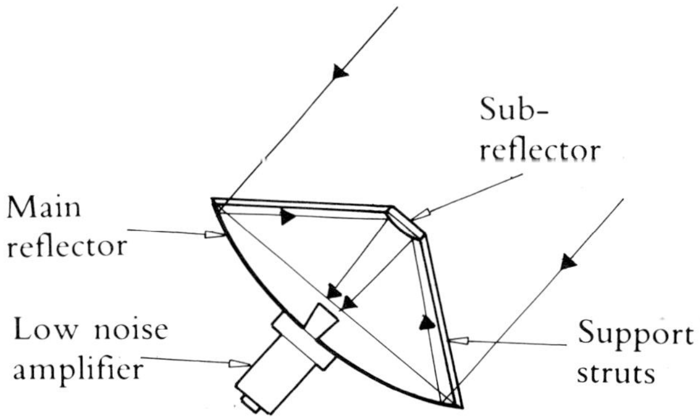

Figure 57 *Some larger stations use a 'Cassegrain' antenna, with a sub-reflector, so that the feed horn may be mounted directly on to the low-noise amplifier, avoiding waveguide loss*

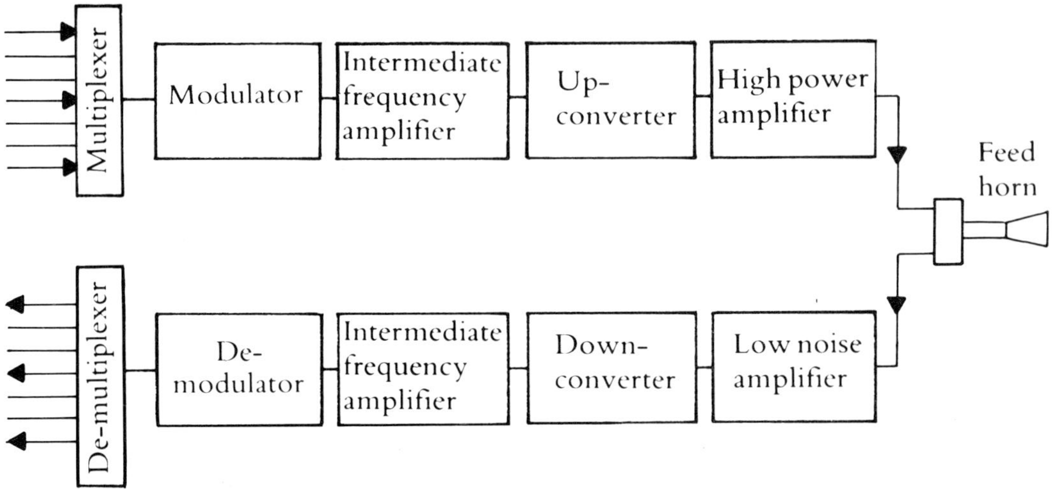

Figure 58 *Principal components of a transmit–receive Earth station*

modulator is fed with an intermediate frequency (IF) carrier, which operates at 70 MHz or 140 MHz depending on the characteristics of the system; this carrier is changed in phase (modulated) according to whether a '1' digit or a '0' is being imposed on it. An IF amplifier then filters and amplifies the phased carrier and passes it to an up-converter which changes the carrier frequency to that of the space system – in this case, 14 GHz. Finally, the phased 14 GHz carrier is passed through a high-power amplifier (HPA) to the antenna for up-linking to the satellite. Depending on the receive-sensitivity of the satellite, the size of the Earth station antenna and the quantity of traffic to be handled, the RF output power of the HPA can vary from about 50 watts to one or two kilowatts.

The receive side operates in reverse. The phased 12 GHz carrier arriving from the satellite is passed first into a low-noise amplifier (LNA) – typically, today, a field-effect transistor (FET) amplifier. The carrier is down-converted to IF, and is filtered and re-amplified at that frequency. After amplification, the signal passes into a demodulator, which detects the phase shifts in the carrier and converts them to the original digits which first imposed the changes at the transmission station. The final digital stream is then demultiplexed into the original voice, data and video signals intended for the receiving station.

Television for entertainment purposes is not yet delivered in digital form. While video links for business purposes can operate satisfactorily at digital bit rates as low as 64 kb/s, high-quality TV needs bit rates of between 90 and 140 Mb/s, over 1000 times faster, and this leads to uneconomic use of satellite transponders. TV is thus still transmitted by frequency modulation (FM) of the up-link and down-link carriers, but the processing of the signals in the Earth station still resembles that of digital processes, except for differences in the modulators and demodulators.

As seen in Chapter 9, the higher the EIRP of the satellite, the smaller the Earth station antenna required to handle a given quantity of traffic. In contrast to today's commercial satellites, which offer EIRPs in the range of 40 to 50 dBW, the satellites of the early 1970s, with lower transponder powers and with wide-area beams such as those needed by Intelsat, provided only 30 to 35 dBW EIRP. To capture enough power to handle hundreds or thousands of telephone circuits, the Earth stations of the 1970s employed very large antennas, some over 30 m in diameter. Because of their large size, their beamwidths were small, sometimes as small as one-tenth of a degree, and this meant that they had to be automatically steered continuously (and still have to be) to ensure that their beams remained 'locked on' to their satellites as they moved within their 'boxes' in the sky. Automatic steering, using radar tracking techniques and drive mechanisms on all three axes of the antenna, is expensive in equipment cost, 24-hour manning, maintenance and spares. It is not at all suited to the economics of TV reception stations for cable distribution systems or for commercial communications networks. What these operations need are Earth stations with antennas small enough to provide wide beams that can keep the satellite in view without having to move the antenna. Once the antenna is pointed at the satellite, it can remain locked in that position permanently.

A fixed antenna is simple, reliable and needs no maintenance. The smaller it is, the lighter it is, and it becomes easier and cheaper to install – on a factory roof or in an office car park, or even indoors, looking out through a window. A typical small commercial station today uses an antenna less than 2 m in diameter.

However, there are limits to smallness. As shown in Chapter 9, the beamwidth of an antenna is inversely proportional to its diameter. As the antenna gets smaller, therefore, there comes a point where the beam is so wide that it can 'see' two satellites – the one being aimed at and another adjacent to it. Today, and in the future, with satellites being placed closer and closer together to make maximum use of the geosynchronous arc, this leads to obvious problems of interference, both in reception and in transmission.

In the section on satellite antennas in Chapter 6, it was seen that 'spill-over' occurs beyond the nominal half-power beamwidth of a satellite down-link beam. The same applies to Earth station antennas, which means that the smaller antenna not only has a wider half-power beamwidth, but it exhibits additional spill-over (known as *sidelobes*) beyond that angle as well. The transmit power and the reception sensitivity in the sidelobes fall off rapidly as the angle widens, but they worsen the interference situation. In determining the size of Earth stations to handle certain traffic requirements with different satellite EIRPs, account has to be taken of these matters of interference. In practice, to avoid exceeding maximum limits of interference between adjacent satellites, the antenna half-power beamwidth should not exceed one-quarter of the spacing between the relevant satellites operating in the various bands. The effect of this empirical rule is shown in Figure 53 (page 82).

Performance

The performance of an Earth station, that is, its traffic-handling capacity with a given satellite EIRP, is dependent on two factors – gain and noise. Antenna gain has been dealt with in Chapter 9. It is measured, as with satellite antennas, in dBi, and is related to antenna diameter and the operating frequency. Noise is a more complex matter. It is generated from a variety of sources, of which the combined effects interfere with the wanted radio signal. As it increases, the signal-to-noise ratio (*S/N*) reduces to a level at which the signal becomes unintelligible.

The principal source of noise occurs within the electronic equipment of the Earth station itself: it is caused by the random motion of electrons, which increases with increasing temperature. For this reason, Earth stations of the 1960s and 1970s used pre-amplifiers cooled by liquid nitrogen to keep internal noise to the absolute minimum, but the increased power of satellites today allows amplifiers to operate at ambient temperatures. The sky contributes a small amount of radio noise, as do radio stars. Because noise is related to temperature, the Earth can contribute noise if the antenna is depressed to a low elevation angle, as it has to be if it is located at high latitudes. In such cases, the effect of the temperature of the Earth can be detected within the antenna beam.

Noise is also generated by rainfall, because the temperature of the rain is higher than that of the clear sky. The major effect of rain is seen in the attenuation of the signal from the satellite as it passes through the rain

A 13 metre (40ft) Cassegrain antenna for tracking, telemetry and command

belt. This is referred to later in Chapter 11, which looks at signal-to-noise ratios in varying weather conditions.

The relationship between noise and temperature in most noise sources leads to the use of the term *noise temperature* (T) in evaluating the performance of Earth stations. The term encompasses all the sources of noise present, including those that are thermally generated and those that are not. Because the equivalent electrical power of thermal noise is proportional to absolute temperature, all the combined noise sources within T are expressed as one value in kelvins (K).

Thus, for example, in a commercial Ku-band station, the FET low-noise amplifier will contribute about 300 K, to which must be added sky noise of about 40 K (depending on elevation angle) and noise from other small sources, such as Earth noise and thermal radiation from the antenna itself and from the waveguide run between the horn and the amplifier. The total noise temperature, T, will amount to about 370 K. The communications performance of the station is then taken as the ratio between the antenna gain (G) and the noise temperature (T) and is shown as G/T, expressed in dB.

Continuing the example above, if the Ku-band station has a 3 m diameter antenna, the formula for gain in Chapter 9 shows that its G at the down-link frequency of 12 GHz will be 49 dBi. The T of 370 K is expressed in dB as 25.7 dBK. Thus the station G/T is $49 - 25.7 = 23.3$ dB.

G/T for a range of antenna diameters, with varying noise temperatures, is shown in Figure 59.

Up-link transmission

Traffic-handling capacity on the up-link, with the Earth station transmitting to the satellite, is dependent on the same three factors that determine down-link capacity – EIRP, gain and noise. In this case, however, it is the EIRP of the Earth station that applies, while gain and noise are those associated with the satellite.

As with the satellite, the EIRP of the Earth station is the product of its antenna gain and its transmitter power. As with the Earth station, the receive performance of the satellite is its G/T, but here it is not possible to increase the antenna diameter to improve performance because that diameter has already been decided by its coverage area (Chapter 9). Further, the antenna is not looking at clear sky, with its fairly low noise temperature of around 40 K, but at the Earth, which emits radiation at nearly 300 K. Thus the receive performance of the satellite is constrained by its relative lack of gain and its higher noise, and is invariably poorer than that of the Earth station working with it. To compensate for this, the Earth station simply uses power in its transmit amplifier. Unlike the satellite, which is limited in the power it can transmit, this higher power in the Earth station causes no supply problems, because the power, usually in the range of a few hundred watts, is simply drawn from the mains. However, because of radiation spill-over that can occur with small antennas, the output RF power of the Earth station may be restricted to prevent interference with adjacent satellites. All stations are indeed subject to international and national regulations on this matter of interference. In some cases, therefore, to provide an adequate up-link EIRP without exceeding regulatory limitations, the gain of the Earth station has to be increased by increasing the antenna diameter, but at the expense of reduced beamwidth and thus the need for more accurate pointing.

Figure 59 (pages 93–94) *Antenna G/T plotted against diameter and system noise temperature for four operating frequencies*

Figure 59(a)
Frequency: 11 GHz

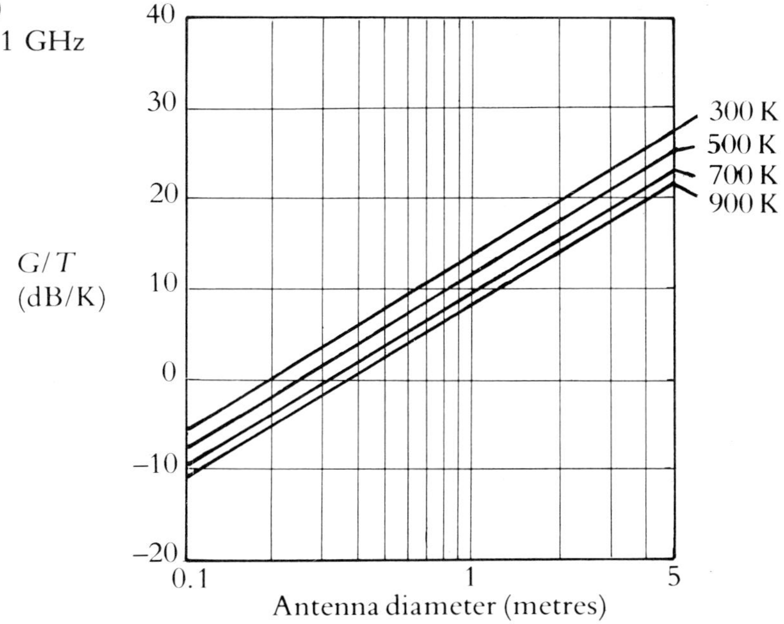

Figure 59(b)
Frequency: 12 GHz

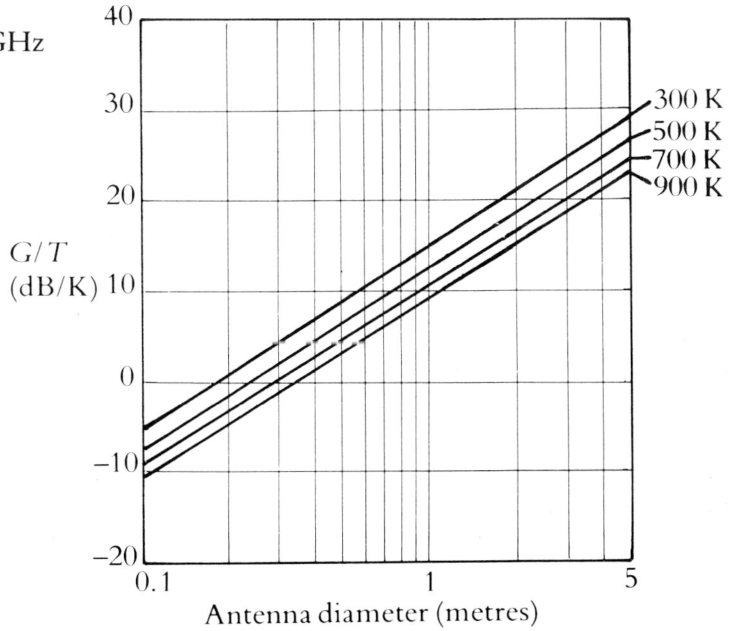

Figure 59(c)
Frequency: 14 GHz

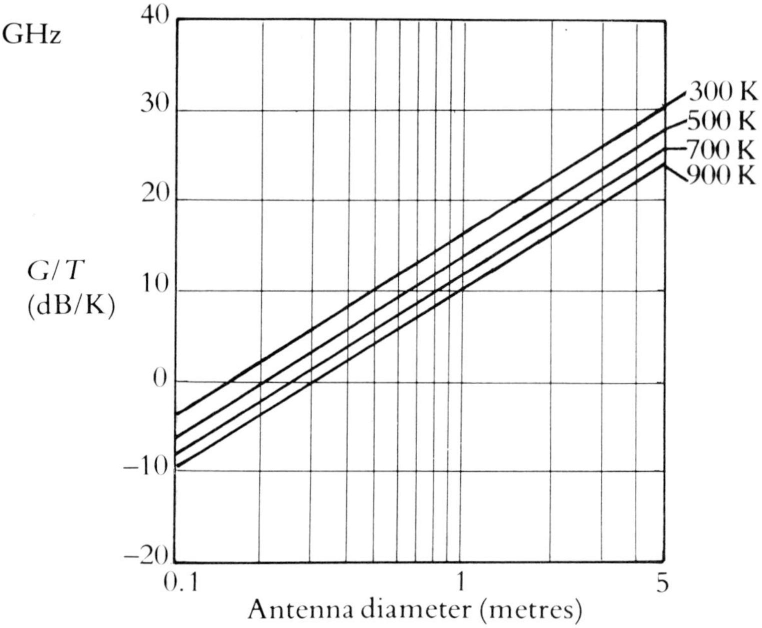

Figure 59(d)
Frequency: 18 GHz

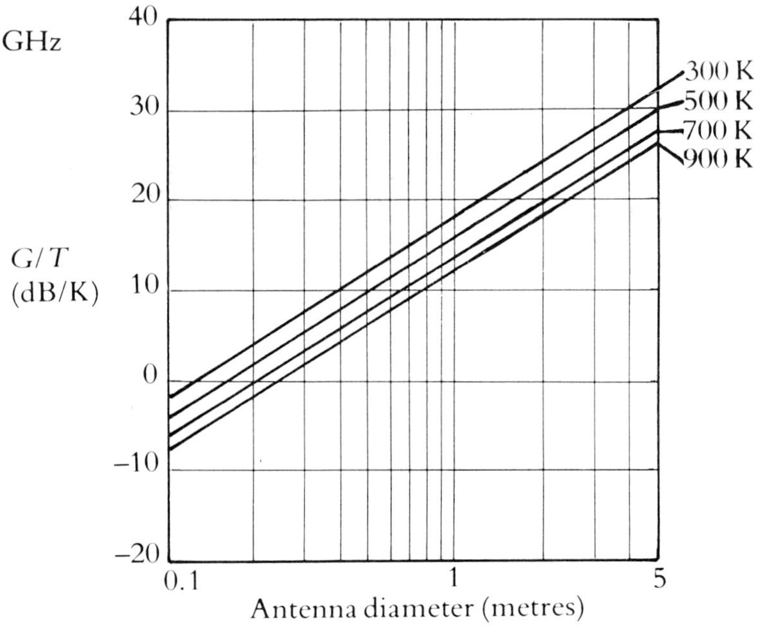

Examples of up-link EIRP requirements, and their importance in determining the overall performance of a satellite link, are given in Chapter 11.

Pointing an Earth station

All operators of commercial communications satellites provide tables to show users where to point their Earth stations to 'find' the satellite that

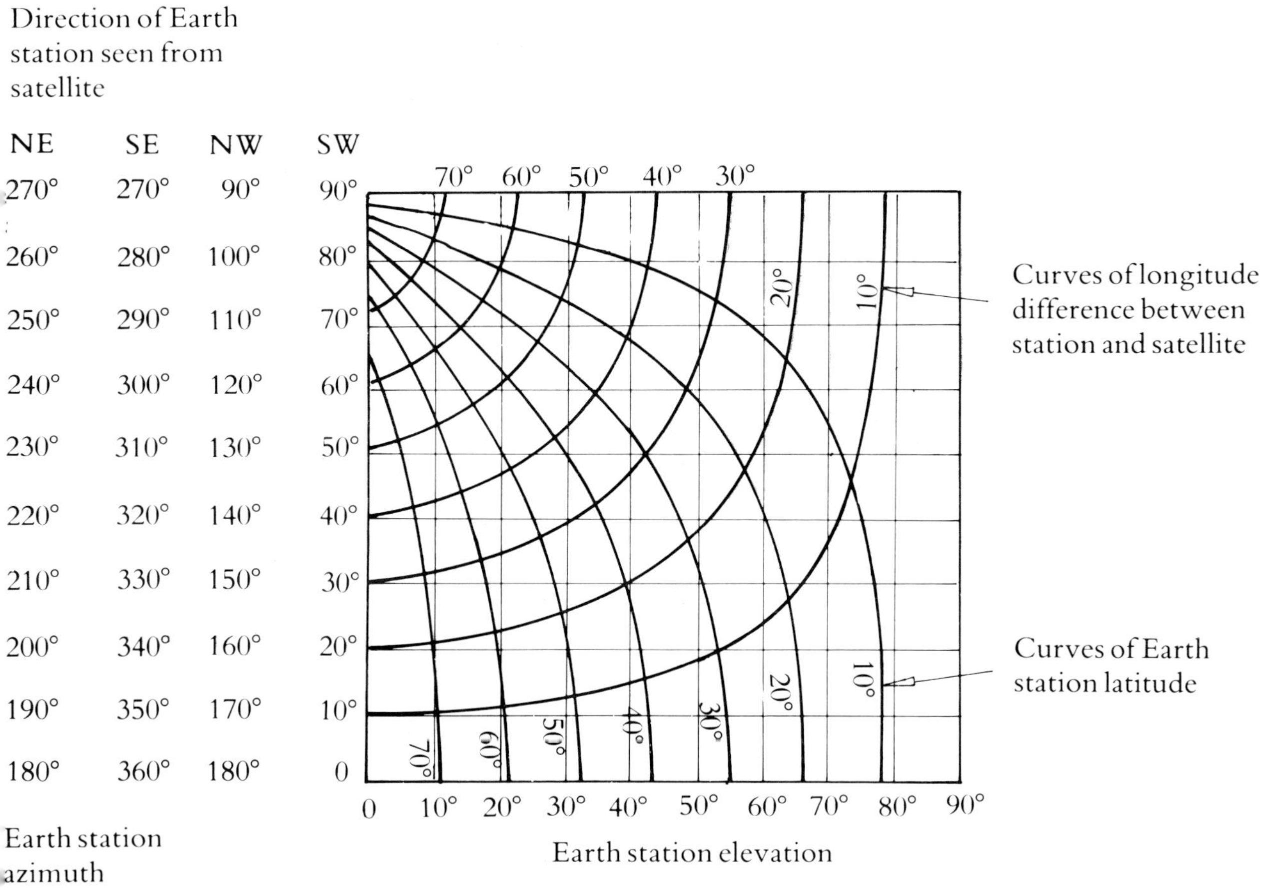

Figure 60 *Earth station pointing angles in elevation and azimuth. Example: Station at Oslo, latitude 60°N, longitude 10°E. Satellite at longitude 30°W. Longitude difference is 40°, so Earth station elevation is 14°. Station is north-east of satellite, so Earth station azimuth is 224° (44° west of south)*
(From a chart produced by Gary Gordon of Comsat Laboratories)

they are going to use. The user locates his exact latitude and longitude in the tables, which then show him the required elevation of the antenna (degrees from horizontal) and azimuth (degrees clockwise from true North).

Figure 60 is a general chart, showing elevation and azimuth angles for any Earth station position and any satellite location in the geosynchronous arc. Here, longitude difference needs to be known, that is, the number of degrees between the longitude of the station and that of the satellite (see the examples in Chapter 15). Generally, a minimum elevation of 5° is satisfactory. Below this, the station can pick up interference from terrestrial radio sources and suffer from Earth heat radiation, which will increase noise in the system.

The chart can clearly offer only an approximation of elevation and azimuth angles, but is good enough for very small Earth stations of about 1 m diameter or less. These can capture the satellite initially in their relatively broad beamwidths, using the chart, and then the antenna can be fine-pointed to receive the best signal. Larger stations need more accurate, tabular pointing data.

Operators' tables should also show when 'Sun-blinding' will occur, that is, when the station, satellite and Sun are in line, which occurs twice per year. The Sun's temperature increases overall noise in the station receiver, which will interfere with communications for a few minutes when this conjunction occurs.

11

Traffic capacity and quality

Preceding chapters have indicated, in a qualitative manner, how traffic capacity and quality are related to EIRP and G/T. This chapter now quantifies those relationships, and shows how the overall performance of a space link can be calculated.

Still using qualitative terms, the overall performance relationships can be shown as an equation:

$$\text{Quality} = \text{EIRP} + G/T - \text{space loss} - \text{a constant} - \text{capacity} - \text{other losses}$$

all expressed in dB.

All these terms, including the newly-appeared 'space loss', 'a constant' (Boltzmann's constant) and 'other losses', now need to be examined and quantified.

Quality

Quality is measured as carrier-to-noise ratio (C/N), expressed in dB. The higher the ratio, the better is the quality of the received signal. Although signal-to-noise ratio (S/N) is used in general communications practice, it is not suited to direct interpretation of quality in space link calculations because, while it is related to C/N, it is affected also by the format and characteristics of the signal itself. It can thus be misleading in the estimation and measurement of the relative qualities of communications links. It is better therefore to confine measurements to C/N, which is not affected by the 'shape' of the signal.

The effect of C/N on quality is seen in two different ways in the two principal uses of space communications: (1) analogue FM television, and (2) digital voice, data and video links used for public and corporate telecommunications.

For FM television, the required value of C/N is wholly subjective. It determines the quantity of noise on the TV screen and is thus subject to individual acceptance of, or objection to that noise, which appears as white spots and streaks all over the picture. A 'perfect', high-quality picture, which looks as good as colour film, without a trace of spots,

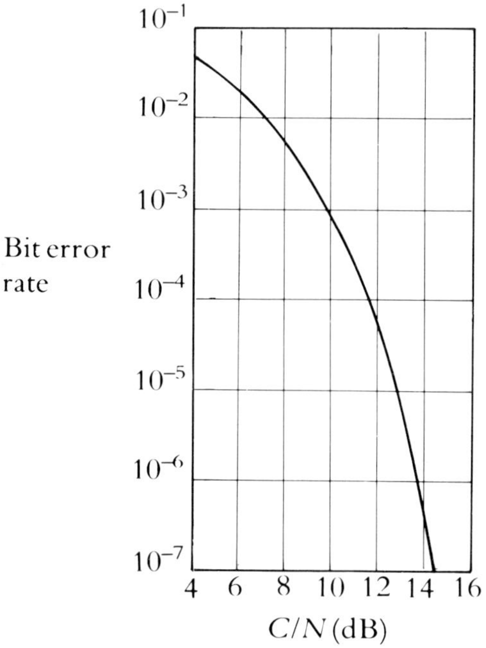

Figure 61 *Bit error rate plotted against C/N for QPSK*

needs a C/N of at least 18 dB (carrier power to noise power is 63:1). Most people are content with $C/N = 14$ dB, with almost imperceptible, occasional white spots, while many viewers today, in areas where there is poor reception from terrestrial transmitters, accept $C/N = 10$ dB without complaint. By the time C/N falls to 6 dB, the picture is very noisy but, here again, some people will accept that quality if that is all they can get and they have a pressing desire to see the programme. Indeed, experiments using new techniques in the encoding of TV have shown that an intelligible picture can be received at C/N as low as 2 dB, but no one would accept this quality for more than a few minutes. The content of the programme also has an effect on subjective acceptability: other experiments have shown that, at $C/N = 6$ dB, the noise is very evident when a standard stationary test card is displayed, but if the picture is full of movement and interest, the noise is hardly noticed.

In the case of digital transmissions for voice, data and video, acceptable C/N is determined more objectively. There is a direct mathematical relationship between C/N and the bit error rate (BER) in the digital signal. This relationship depends on how the carrier wave is modulated by the digital signal. An example, where the phase of the carrier is modulated four times per cycle by the '1' and '0' digits imposed on it, is shown in Figure 61.

Bit error rate exhibits most impact, of course, on data communications, particularly in computer-to-computer links. Many error-detection and error-correction systems exist to combat the effects of noise in data circuits, but even with these a BER better than 1 error in 1 million bits (BER $= 10^{-6}$) is usually demanded. As Figure 61 shows, this requires a C/N of at least 14 dB. In practice, 16 dB is usually applied. Digital voice and commercial video, such as that used for teleconferencing and for the viewing of documents and equipment, can accept a lower C/N, in the range 10 to 12 dB.

EIRP and *G/T*

EIRP and G/T, the two positive items in the communications equation, have been dealt with in Chapters 9 and 10. The higher they are, the better the C/N.

Space loss

Space loss comprises two components. The first is called *spreading loss* and simply represents the surface area of a sphere of radius 36 000 km (22 000 miles) over which the satellite signal would be spread by the time it reached the Earth if there were no antenna to capture and focus it (see Chapter 9). The same spreading loss applies to the up-link from an Earth station (again, if no antenna were fitted). The value of this first component is approximately 163 dB, depending on the exact distance from the satellite to the Earth station.

The second component is related to frequency, and is measured as $-(\lambda^2/4\pi)$. At a frequency of 12 GHz, $\lambda = 0.025$ m (Chapter 8), so this component equals $-0.000\,05$, which is $+43$ dB. (Two minuses produce the plus sign.)

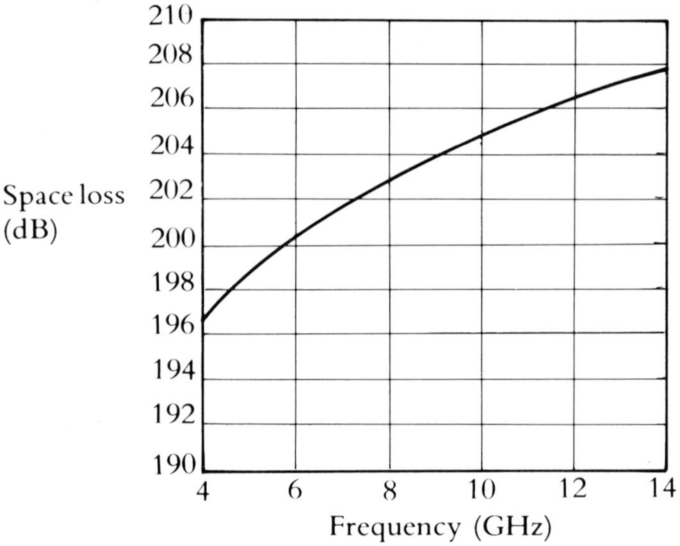

Figure 62 *Space loss plotted against operating frequency*

The two components are added together to give total space loss, which in this case becomes 163 + 43 = 206 dB.

Figure 62 shows space loss for a range of operating frequencies.

Boltzmann's constant

Boltzmann's constant is derived from the Maxwell–Boltzmann principle of physics that is applied to the thermal energy of particles and electrons. It states that the relationship between the energy of a particle and its motion is related to its absolute temperature multiplied by Boltzmann's constant, which has the value of −228.6 dB. This very small factor ($10^{-22.86}$ joules per kelvin) exerts a powerful influence on the effects of noise and thus on quality. Because it is a negative number, its value is *added* to the communications equation when it is deducted as a constant.

Capacity

Capacity is measured in terms of *bandwidth*, again in dB. Bandwidth is the number of cycles per second (hertz) required to carry an analogue signal. If the signal is in digital form, the equivalent term is *data rate*, measured in bits per second (b/s). As shown in Chapter 8, Hz and b/s are prefixed by kilo, mega or giga to show units in thousands, thus:

1 kHz	= 1000 Hz	1 kb/s	= 1000 b/s
1 MHz	= 1000 kHz	1 Mb/s	= 1000 kb/s
1 GHz	= 1000 MHz	1 Gb/s	= 1000 Mb/s

In analogue circuits, bandwidths range from 4 kHz for a voice channel to 5.5 MHz for 625-line colour TV, which needs a carrier bandwidth of 27 MHz. With digital circuits, voice can be accommodated in 16 kb/s, although 32 kb/s is more common. Typical data links run at 9.6 kb/s, while video for teleconferencing can range from 2 Mb/s down to 56 kb/s depending on the picture content and movement in the picture. Colour TV of 625-line entertainment quality requires between 70 and 90 Mb/s, while future high-definition TV (HDTV), operating at 1125 lines, will require a bandwidth of 144 Mb/s.

The total bandwidth of a space communications link is the bandwidth of all the individual signals multiplexed into it, added together. Thus, the total bandwidth of a link carrying 1000 analogue voice channels, each operating at 4 kHz, would be 4 MHz, plus an allowance to avoid adjacent channels interfering with each other. A multiplexed link of 100 digital voice channels, each running at 32 kb/s, plus 10 videoconference channels, each at 2 Mb/s, would require a total bandwidth of 23.2 Mb/s.

The communications equation calls for the bandwidth to be expressed in dB. In addition, because most of the other terms in the equation are related to frequency, bandwidth must be too, and be shown in dB Hz. In the 1000 analogue voice circuit example, the bandwidth of 4 MHz would appear as 66 dB Hz. In the digital case, the data rate must be converted to an equivalent analogue bandwidth.

The conversion is based on the manner in which the digital signal is imposed on the analogue carrier wave to modulate it. One of the most common practices is to use QPSK (quaternary phase-shift keying), in which the phase of the carrier wave is changed by the application of pairs of digits four times per cycle. In this case, the working figure for the digital capacity of the analogue carrier is 1.5 bits per hertz. Thus, the example above of 23.2 Mb/s would need a bandwidth of 15.5 MHz, which would appear in the equation as 71.9 dB Hz.

The digital capacity of an analogue wave can be increased by applying the digit inputs more times per cycle. For example, with 8-PSK (eight times per cycle) the capacity rises to around 4 bits per hertz, and the required analogue bandwidth to carry a given data rate falls accordingly.

Splitting each cycle into finer sectors in this way requires higher-quality modulation equipment, which is more expensive than QPSK, and the modulator is more susceptible to angular 'jitter' which arises from noise in the system.

It is because of noise that bandwidth is referred to as 'noise bandwidth' in the communications equation. Noise, generated by external sources, and by the internal electronics of all the equipment used in the system, is directly proportional to the bandwidth of the circuit, which is why noise bandwidth carries a negative sign in the equation. The higher the bandwidth, the greater the noise, and the quality suffers accordingly.

Other losses

In estimating the performance of a satellite link by using the communications equation, account has to be taken of other losses that appear in addition to space loss. The principal loss, caused by rainfall, varies with the intensity of the rain and with the operating frequency. Figure 50 (page 76) shows examples of the range of losses that must be tolerated under varying rain conditions at different frequencies. It shows, for example, that at 12 GHz, a moderate rain rate of 4 mm/hour will attenuate the signal at a rate of about 0.1 dB per kilometre; with very heavy rain falling at 50 mm/hour, the loss rises to around 3 dB per kilometre. The number of kilometres is that of the slanting distance of the signal as it passes through the rain belt. For estimation purposes, the total loss is typically assumed to lie between 5 and 10 dB, account being taken of the number of hours or minutes per year over which that loss is suffered.

Other assumed and measured losses are those arising from various atmospheric effects, and are usually taken to lie between 1 and 2 dB.

Estimating traffic capacity and quality

With all the terms of the communications equation explained and quantified, it is now possible to estimate the value of any one of the terms, given all the others.

Repeating the equation and inserting Boltzmann's constant and the quantifiable names for quality and capacity, it becomes

$$C/N = \text{EIRP} + G/T - \text{space loss} + 228.6\ \text{dB} - \text{bandwidth} - \text{other losses}$$

Example estimations can now be undertaken.

Example 1

A direct-to-the-home television broadcast system is to be established, working into home antennas of 60 cm (2 ft) diameter. The system will provide 625-line colour TV, needing 27 MHz of bandwidth. The satellite will offer an EIRP of 60 dBW, operating at a downlink frequency of 12 GHz. What picture quality will be achieved?

For quality, *C/N* needs to be determined:

EIRP	60.0 dBW	
plus *G/T*	+ 8.0 dB/K	(From Figure 59(b) at 500 K)
minus space loss	− 206.4 dB	(From Figure 62 at 12 GHz)
plus Boltzmann	+ 228.6 dB	
minus bandwidth	− 74.3 dB Hz	(27 MHz)
minus other losses	− 1.0 dB	
C/N	= 14.9 dB	in clear sky conditions
minus rain loss	− 5.0 dB	
C/N	= 9.9 dB	in heavy rain conditions

It has been seen, under 'Quality' in this chapter, that a *C/N* of 14.9 dB will provide a good, though not perfect picture. In heavy rain, however, the picture quality falls to a lower but acceptable level, and this may occur for only a few minutes per year.

Doubling the *area* of the home antenna, that is, increasing its gain by a factor of 2, will increase its *G/T* by 3 dB, and thus the *C/N* by 3 dB. Thus, an antenna of 85 cm diameter (twice the area of a 60 cm) would permit adequate viewing in heavy rain, and perfect quality (17.9 dB) in good weather. To obtain these conditions with the 60 cm antenna, the satellite power would need to be doubled, from 60 to 63 dBW.

Example 2

What EIRP is required to provide a television distribution system into 3 m diameter antennas owned by cable distribution operators, given the following conditions?

Required *C/N* = 14 dB   Bandwidth = 27 MHz
Frequency = 12 MHz

In this case, it is EIRP that is being sought. This term has to be taken to the left-hand side of the equation, and C/N taken to the right. When moving a term from one side of the equation to the other, its positive or negative sign must be changed, so the equation first becomes

$$-\text{EIRP} = -C/N + G/T - \text{space loss} + 228.6\ \text{dB} - \text{bandwidth} - \text{other losses}$$

But EIRP is required as a positive number, so *all* the signs in the equation need to be changed, giving

$$\text{EIRP} = C/N - G/T + \text{space loss} - 228.6\ \text{dB} + \text{bandwidth} + \text{other losses}$$

C/N	14.0 dB	
$-\ G/T$	$-$ 23.0 dB/K	(From Figure 59(b) at 400 K)
+ space loss	+206.4 dB	(From Figure 62 at 12 GHz)
− Boltzmann	−228.6 dB	
+ bandwidth	+ 74.3 dB Hz	(27 MHz)
+ other losses	+ 1.0 dB	
EIRP	= 44.1 dBW	in clear sky conditions

Here again, allowing for a 5 dB rain margin, the satellite power would need to be increased to 49.1 dBW.

Example 3

What traffic capacity can be handled by a 2.5 m Earth station operating in a commercial communications network at a down-link frequency of 11 GHz, with a satellite providing an EIRP of 45 dBW? For a BER better than 1 in 10^6, a C/N of 16 dB is required.

In this case, bandwidth needs to be calculated. Moving the equation terms and transposing + and − signs as in Example 2, the equation becomes

$$\text{Bandwidth} = \text{EIRP} + G/T - \text{space loss} + 228.6\ \text{dB} - C/N - \text{other losses}$$

EIRP	45.0 dBW	
+ G/T	+ 20.5 dB/K	(From Figure 59(a) at 400 K)
− space loss	− 205.5 dB	(From Figure 62 at 11 GHz)
+ Boltzmann	+ 228.6 dB	
− C/N	− 16.0 dB	
− other losses	− 2.0 dB	
Bandwidth	= 70.6 dB Hz	= 11.5 MHz in clear sky conditions

In digital terms, taking the QPSK ratio of 1.5 bits per hertz, this equates to 17.2 Mb/s. With a 5 dB rain loss, this falls to 5.5 Mb/s.

These three examples have been applied to down-links only. The communications equation works in exactly the same way on the up-link, but in this case EIRP is that of the Earth station and G/T is that of the satellite. The G/T of the satellite is fixed because the diameter of its antenna has already been determined by the coverage area, with a T higher than that of a sky-pointing Earth station (Chapter 10). Example 4 shows the determination of required up-link transmitter power.

Example 4

What RF transmitter power is required on a 3 m Earth station to up-link a bandwidth of 36 MHz at 14 GHz with a C/N of 14 dB? The satellite carries an antenna of 0.5 m diameter, with a system noise temperature (T) of 750 K.

Earth station EIRP is needed first, so the equation of Example 2 can be used:

C/N	14.0 dB	
− G/T (satellite)	− 6.0 dB/K	(From Figure 59(c) at 750 K)
+ space loss	+ 207.5 dB	(From Figure 62 at 14 GHz)
− Boltzmann	− 228.6 dB	
+ bandwidth	+ 75.6 dB Hz	(36 MHz)
+ other losses	+ 2.0 dB	
+ rain loss	+ 10.0 dB	
Earth station EIRP	= 74.5 dBW	

The resulting EIRP of 74.5 dBW is the product of the gain of the Earth station antenna and its transmitter power. The gain of a 3 m antenna at 14 GHz is 50 dB (Figure 52) so the transmitter must produce 24.5 dBW, which is 280 watts. In practice, a standard 500 watt transmitter would be used, 'backed-off' to provide the required up-link *C/N* in normal conditions, with an ability to increase power in extremely adverse weather conditions.

Example 5 – Overall C/N

If the *C/N* on the up-link is different from that on the down-link (which is usual), the overall *C/N* of the entire link, from the transmitting station via the satellite to the receiving station, is found from the formula

$$\frac{1}{C/N \text{ overall}} = \frac{1}{C/N \text{ up}} + \frac{1}{C/N \text{ down}}$$

Here, the *C/N* values have to be converted to arithmetic numbers from dB, and then converted back again. For example:

$$C/N \text{ up} = 14 \text{ dB} = 25.1$$
$$C/N \text{ down} = 16 \text{ dB} = 39.8$$

Then

$$\frac{1}{C/N \text{ overall}} = \frac{1}{25.1} + \frac{1}{39.8}$$

$$= 0.39 + 0.025 = 0.064$$

So $C/N \text{ overall} = 1/0.064 = 15.6 = 11.9 \text{ dB}$

This shows the importance, when 'up' and 'down' *C/N*s differ by only a few dB, of keeping their values as close together as possible. In many cases, however, close matching of 'up' and 'down' *C/N*s is not practical; it then becomes necessary to make one *C/N* value (either the 'up' or the 'down') very much higher than the other. This minimises the loss in overall *C/N*, as seen in the following example.

Example 6

A direct-broadcast TV system is designed to provide a *C/N* of 16 dB into 60 cm home antennas. What will be the overall *C/N* if the up-link transmitter station, employing a 13 m antenna, provides an up-link C/N of 30 dB?

$$C/N \text{ up} = 30 \text{ dB} = 1000$$
$$C/N \text{ down} = 16 \text{ dB} = 39.8$$

Then
$$\frac{1}{C/N \text{ overall}} = \frac{1}{1000} + \frac{1}{39.8}$$
$$= 0.001 + 0.025 = 0.026$$

So
$$C/N \text{ overall} = 1/0.026 = 38.5 = 15.86 \text{ dB}$$

The overall loss is thus only 0.14 dB.

Similar situations apply in systems for mobile users where, to obtain a reasonable overall C/N through the very small stations carried on ships, aircraft and trucks, the signals are routed into and out of large, powerful fixed stations that provide very high C/Ns in relation to those of the mobile users.

Example 7 – Multiple users

In Example 3, traffic capacity for a single Earth station was examined. What is the effect of several Earth stations sharing a transponder?

The effect is dependent on the method of access to the transponder, which is explained in Chapter 13. If frequency division access is used, each station is allocated a section of the overall bandwidth of the transponder, and the carrier power (C) to each station is reduced proportionately. Thus, with 10 stations sharing one transponder, the C to each station is reduced to one-tenth. But, because each station is using only one-tenth of the transponder bandwidth, its noise (N) is reduced by a factor of ten as well, so the quality (C/N) of the received signal over the smaller bandwidth remains unchanged. In practice, to maintain linearity in the transponder, that is, to ensure that it does not distort the many adjacent signals that are passing through it, its output power has to be reduced by up to 6 dB. The bandwidth available to each station, or the number of sharing stations, must therefore be reduced accordingly.

This constraint does not apply if time division access is employed but, as explained in Chapter 13, this form of access calls for much more power and complex equipment in each Earth station.

12

Selecting a satellite

The previous chapter has shown that EIRP, C/N, G/T and bandwidth are the decisive variables in the communications equation. EIRP, being the product of two factors – antenna gain and transmitter power – allows the system designer to estimate the capacity and quality of traffic without having to worry about how that EIRP is achieved. As explained in Chapter 9, it may be derived from a very powerful satellite with a small antenna, or from a very low-powered satellite with a large antenna. When the system designer comes to select a suitable satellite to provide that EIRP, he must first determine coverage area because that will decide antenna size. Once this is known, he can then select his satellite on the basis of its capability in power and in bandwidth.

For example, take again Example 1 of the previous chapter, where it is seen that an EIRP of 63 dBW is required to provide a good-quality colour TV picture via 60 cm home antennas in bad weather conditions. Assume that the coverage area is the British Isles. This is best covered by an elliptical antenna to give an elliptical coverage, as seen in Figure 41 (page 60), but for simplicity here a circular antenna can be assumed, providing a circular 'footprint' of about 1000 km diameter. From the beamwidth formula given in Chapter 9, it will be seen that this footprint will require a beamwidth of about 1.75° and thus, at the 12 GHz down-link operating frequency, a diameter of 1 metre. The gain of a 1 m antenna at 12 GHz (again from Chapter 9) is 39.5 dB. Thus the output RF power of the satellite transponder, *per TV channel*, must be $63 - 39.5 = 23.5$ dBW, which is 225 watts.

If three TV programmes are required by the system, the satellite must provide $3 \times 225 = 675$ RF watts; for five channels, 1125 RF watts. The question then arises – what satellites can provide these levels of power?

The question is best answered by taking an example range of communications satellites manufactured by one of the world's suppliers – in this case British Aerospace and its Anglo-French arm, British Aerospace-Matra.

The range can be seen in Figure 63. Three classes are shown: ECS, which is employed for telecommunications and low-power TV distribution; *Eurostar*, which is going into service for maritime communications and higher-power services; and *Olympus*, aimed particularly at high-power television broadcasting. The figure shows the maximum RF power that can be developed by each class of satellite, and from this it is clear that

	Maximum DC input power (watts)	*Maximum RF output power* (watts)	*Maximum mass at launch* (kg)	*Maximum mass of communications payload* (kg)
ECS	1000	300	1000	100
Eurostar	2500	1000	2000	350
Olympus	7500	2500	3300	650

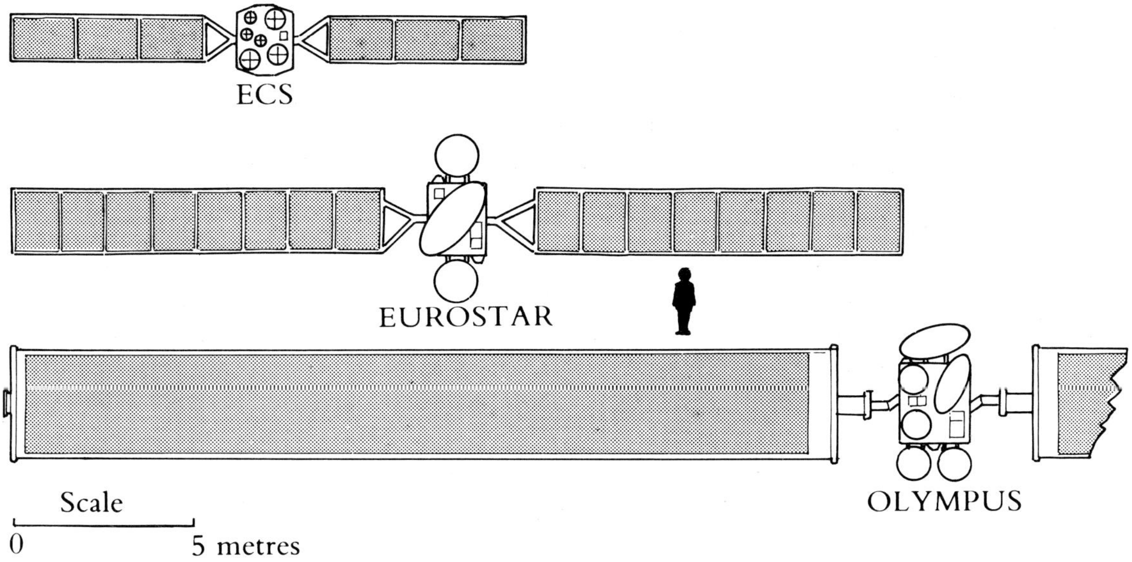

Figure 63 *British Aeropace satellite classes. In* ECS *and* Eurostar*, the joints in the solar arrays, which fold up concertina-fashion for stowage in the launcher, can be seen. With* Olympus*, the solar array cells are mounted on a flexible blanket, which is folded for launch and then stretched out by an erecting mast behind each array*

the ECS class would not be suited to the example, because it could handle only one 225 watt TV channel. *Eurostar* could handle four channels, while *Olympus*, in its maximum power form, could offer eleven. *Olympus*, being more expensive than *Eurostar*, and being a lot heavier, which would lead to higher launching costs, would clearly not be an economic solution in this case. *Eurostar* is the obvious choice.

Eurostar would also be satisfactory from the point of view of bandwidth. The common bandwidth of high-power satellite transponders is 36 MHz, so one transponder would be required for each TV channel. The volume and mass of each transponder is such that *Eurostar* can accommodate up to eight of them. It can thus provide four active channels and four standby channels to be switched on in the event of failure. *Olympus* can accommodate sixteen transponders of this type, giving it the ability to provide, say, 10 active channels with six on standby.

The matter of satellite bandwidth becomes as important as satellite power when telecommunications applications are considered. This is illustrated by reference to Example 3 of the previous chapter. Here, an EIRP of 45 dBW is required for a system using 2.5 metre Earth stations in a corporate communications network. Once again, account must first be taken of coverage area to determine satellite transponder power. Taking that the coverage area is to be most of Europe, it will be found, as a first approximation, that this will require a 4° beam from the satellite. The antenna diameter to give 4° at the specified down-link frequency of 11 GHz is 0.5 m. This provides a gain of 32 dB, so transponder power must be $45 - 32 = 13$ dBW, which is 20 watts. From Figure 63, it can be seen that an ECS-class satellite could accommodate, from the point of view of power alone, fifteen such transponders. In practice, because of limitations in internal volume and mass-carrying capacity, it is limited to twelve. Again, on power alone, *Eurostar* could handle fifty 20 watt transponders, but is limited in practice, for the same reasons, to thirty, while *Olympus* is limited to about sixty.

With each individual transponder having a 36 MHz bandwidth, it can be seen that it may be bandwidth, rather than power, that determines the choice of satellite. But this is finally dependent on the bandwidth requirements of the whole network. In the example chosen, it is seen that, in clear sky conditions, the bandwidth that can be handled by each of the 2 m Earth stations is 23 MHz, or 34 Mb/s of mixed voice, data and video circuits. Taking into account signal separation to minimise interference between adjacent signals, this bandwidth will entirely fill the capacity of one 36 MHz transponder on the satellite. If there were thirty stations in the corporate network, all operating simultaneously at maximum traffic rate (outbound and inbound), thirty transponders would be required, and a *Eurostar*-class satellite would be needed. An ECS-class satellite could accommodate only twelve, even though it could provide power for about fifteen 20 watt amplifiers.

In practice, of course, no network operates at maximum capacity all the time. It is here that traffic analyses have to be undertaken, looking at the content, pattern and timing of the traffic in the network, and using traffic-queueing theory to predict waiting times in peak traffic hours against available bandwidth. Such analyses can indicate that, in practice, the required capacity will be less than one-third of the theoretical maximum rate, so that, in this case, the network could be handled by ten transponders in an ECS satellite.

13

Economic system optimisation

The system designer's task is to provide the most economical solution to a given need. He must look for the cheapest satellite that will provide the required power and bandwidth; he must ensure that its launching costs will not be excessive because of weight or size or complexity; he must seek the smallest, simplest, cheapest Earth stations that will offer the required G/T; and he must balance all these against the variations that he can offer in traffic quality and capacity.

In addition to considering all these variables, which have been discussed in previous chapters, he has to take account of the many ways in which the overall capacity of the network can be enhanced – to get more traffic through a limited bandwidth; to enable more Earth stations to be accommodated within a system; to provide higher quality without sacrificing capacity, and *vice versa* – all of which are aimed at increasing profitability for the owner who has invested in a communications satellite system.

Some of these ways of enhancing system performance are described in this chapter. Except for access methods, which are specific to satellite communications, the descriptions of other techniques are necessarily brief because they properly belong to the world of general telecommunications. They are treated fully in relevant textbooks. But all of them, and others, must be considered by the space system designer.

Access methods

In the previous chapter, the problem of trading satellite power against bandwidth was examined, emphasising the need to analyse network traffic patterns in detail to optimise the number of transponders required.

These traffic analyses need to take account of the methods used to give users access to the satellite, so that many of them can operate in parallel through a single transponder without interfering with each other. Two principal methods are in use today, with a third coming into widening operation, and each has its technical and economic advantages and disadvantages.

The first method is called FDMA – Frequency Division Multiple Access – in which the bandwidth of each transponder is subdivided into sections of frequency. Each Earth station in the network is assigned one or more of these sections, for up-linking and down-linking, according to its predicted traffic demands.

The advantage of FDMA is its simplicity and cheapness for the user. Its principal disadvantage is that the transponder has to be 'backed off', that is, run at less than its nominal maximum power, to retain linearity in its amplification response while avoiding intermodulation between the closely-spaced signals that are passing through it. A further disadvantage of FDMA is that frequency assignments cannot be changed instantaneously – a facility known as *demand assignment* – without additional, expensive signalling equipment. As a result, the transponders are likely to be under-utilised on some assignments, while others are causing traffic bottlenecks.

The second method, called TDMA – Time Division Multiple Access – employs a high-speed timing control throughout the network, in which each Earth station transmits its signals in short 'bursts', a few milliseconds long, in a sequence established by a central control station. The sequence itself is variable, changing instantaneously according to the traffic demands of the stations in the network. Each burst, which is initiated from the control station via the satellite, contains address information in advance of the message, so that the message is detected and demodulated only by the station or stations with that address.

The significant advantage of TDMA over FDMA is that each sequential burst, which can use the entire bandwidth of the transponder because only that burst is passing through it at the time, allows the transponder to operate at 'saturation', that is, full nominal power. This can enhance the traffic capacity of the transponder by a factor of two in relation to FDMA.

TDMA operates only with digitised signals, but this offers additional flexibility in that various types of traffic (digitised voice, data and video) can be mixed in a digital stream, and can be addressed and accessed without mutual interference.

The main disadvantage of TDMA is its complexity and cost, with every station in the network having to be equipped with precise timing devices controlled through the satellite. However, cost will inevitably decline with time and with the increasing application of dedicated microprocessors to the timing and control systems. A further disadvantage lies in the need for the Earth station to operate at peak transmitter power to saturate the transponder during each burst.

A third method, which has been developed from military applications, and which will come into increasing use in commercial systems, is called CDMA – Code Division Multiple Access. Here, digital streams are modified by computer-generated 'pseudo-random codes'. All the streams are transmitted together, using the entire bandwidth of the transponder, that is, spread across its entire spectrum of frequency. This gives rise to the term *spread spectrum* that is sometimes applied to this method. Mutual interference between the digital streams is negligible because billions of different codes can be produced economically by today's microprocessors, which also act as economical decoders. The advantages of CDMA are, like TDMA, optimum use of transponder bandwidth and power; intrinsic message security; and low interference levels, because the signal energy is spread over a wide bandwidth, which permits the use of Earth stations that are smaller than those needed in equivalent TDMA and FDMA systems.

Both FDMA and TDMA can be used in ways different from those just described. SCPC (Single Channel Per Carrier) is a form of FDMA in which each carrier frequency assigned to Earth stations carries only one channel (unlike full FDMA, where many channels are multiplexed on to a single carrier). SCPC is particularly well suited to low-density traffic routes in developing countries because of the extreme simplicity of the equipment. SS-TDMA (Satellite-Switched TDMA) is an improvement on TDMA in which addressed messages are switched in the satellite to appropriate down-link beams, and is particularly applicable to multi-beam coverage of service areas. SS-TDMA can also take advantage of regeneration techniques, which improve the overall quality of a link. Regeneration involves accepting the up-link signals in the satellite and converting them down to the format of the original voice, data and video signals – the 'baseband' signals – and then switching them in that format before up-converting them again to the down-link transmission frequency. In this way, bit error rates in the overall link are improved because the relationship between BER and C/N (Chapter 11) applies separately to the up-link and the down-link C/Ns rather than to the overall C/N.

Modulation techniques

Traditional frequency modulation (FM) of baseband signals is being steadily replaced, in both terrestrial and space systems, by digital techniques in which originating analogue waveforms are converted to 1 and 0 binary digits (bits) before transmission. Data signals for computers, facsimile and video text are, of course, already in digital form when originated. Analogue voice, sound and video is converted to digital form by Pulse Code Modulation (PCM) by which the waveform is divided into discrete steps at various rates. Each step is given a digital value, usually of four bits, and these bits are then applied to the carrier wave to modulate it.

As mentioned in Chapter 11 under 'Capacity', the most common way of modulating the carrier wave is called *phase key shifting* (PSK). This advances or retards the phase of the wave according to whether a 1 or a 0 has been imposed on it. When this phase-shifting is applied twice per cycle, the modulation is referred to as BPSK (Binary PSK); when four times, QPSK (Quaternary PSK). At the receiving end, a PSK demodulator detects the phase changes and converts them back to the original bits.

PCM-PSK offers significant advantages over FM. It is far more tolerant of noise, and it permits various types of traffic – voice, data and video – to be mixed into a single data stream for transmission. It can also be manipulated by microprocessors to save bandwidth, reduce power, and offer high security. Some of these manipulations are described below.

Compression techniques

Techniques applied to digital circuits to reduce their bandwidth while retaining signal quality are grouped under the term *compression*. One well-used version is Digital Speech Interpolation (DSI), which takes advantage of the inherent redundancy in the human voice to reduce bandwidth required by a factor of two.

Another is delta modulation, with which transmit power is used only when the digit stream shows a change from 0 to 1, or *vice versa*. If there is a stream of 1s in the message, the transmitter cuts power after the first 1 and stays off until a 0 appears. In the meantime, the receiver 'reads' continuous 1s until it sees the first 0; the same process then repeats with the 0s. All this is happening, of course, at rates of megabits or kilobits per second, and the effect is again the doubling of the capacity of a single circuit.

The ever-growing capabilities of microprocessors have brought into use other forms of manipulation that would not have been economical a few years ago. They can undertake instantaneous analysis of a digital stream and predict the format of subsequent bit groups, providing what is known as predictive coding. With this technique, the bandwidth of a speech signal can be reduced by a factor of four, without noticeable effect on quality. With digital video signals, such as those used for teleconferencing, in which much of the picture remains stationary for most of the time, the bandwidth can be reduced by a ratio of over 30 to 1.

Encryption

Digital circuits are particularly well suited to the application of encryption techniques which, like predictive coding, have become economically practical with the advent of microprocessors. They are as powerful as yesterday's big computers, which could be afforded only by military establishments for their secret messages. Today, encryption and decryption devices for each end of a digital circuit, capable of handling data rates up to 2 Mb/s, can be bought for the price of a good word processor. As prices continue to fall they are finding increasing application in the world of commercial communications.

The way encryption works is explained in Chapter 14.

Echo and delay

Echo occurs in all telephone circuits, arising from the reflection of the speaker's voice in the equipment that is used to convert long-distance '4-wire' links into local '2-wire' distribution networks. It is normally not noticeable because the voice signals are travelling at about 124 000 miles per second ($\frac{2}{3}$ × the speed of light). Even in a long-distance call of a few thousand miles, the echo, which is usually reduced by electronic means, returns within a few hundredths of a second and cannot be detected by the human ear.

With satellite circuits, however, echo was much more noticeable until echo cancellers were introduced. This is because the voice signals have to travel over 22 000 miles up to the satellite, then 22 000 miles down again, followed by a return journey of 44 000 miles. This round trip of nearly 90 000 miles caused echos to return nearly a half-second later, which caused much disturbance to early users. Echo cancellers, now being installed on most satellite circuits, detect the outgoing signal, generate an inverse of it, and then apply that inverse to the line at the exact time the echo is expected. The net effect is total cancellation of the echo.

Regrettably, no solution has yet been found to the problem of the half-second delay itself. When two people are using a terrestrial telephone, each expects to hear an instantaneous response from the other, interspersed with the usual 'uh-huhs', 'mmms' etc that indicate that the listener is indeed listening and understanding. But when the speaker stops speaking and nothing happens for a full half-second, he gets the impression that he has lost contact with his listener, so he starts to repeat his statement. At that moment, the beginning of the listener's response arrives, while he in turn receives the beginning of the repeat. Chaos ensues, until both learn to wait for that half-second; this is very difficult for those used to the cut and thrust of a fast conversation.

One way out of this difficulty is to adopt the practice of those in the maritime and aeronautical worlds who still use one-way radio-telephones and who are trained to say 'over' at the end of each statement, clearing the line for the return response. If, when using a satellite circuit, you mouth silently the word 'over' (which takes about a half-second), you will find that the response arrives when you expect it. But it is difficult to envisage, say, two stockbrokers or two lawyers silently mouthing 'over' at the end of each statement, and it is clear that delay will adversely affect the popularity of voice communications via satellite.

Perhaps not surprisingly, the addition of video to voice negates the problem. Even though the delay is still there, the speaker can literally see the listener accepting his words, and he is quite willing to wait until the listener opens his mouth in response. But while voice can be compressed into 16 kb/s, video needs a minimum additional bandwidth of 56 kb/s, so there is a price to pay for this solution.

14

Encryption

Concern is often expressed by potential users of commercial space systems about message security. It is true that any message transmitted through a satellite can be received by anyone with a suitable Earth station in the coverage area (indeed, all forms of communications, on Earth or in space, can be tapped). However, inexpensive digital encryption systems are now widely available. These systems, based on microprocessors designed specifically for encryption and decryption, have brought to commercial communications levels of security that were un-dreamed of a few years ago. These levels match those of many military systems that are still in use today, but they are available at a tiny fraction of the cost of those systems – with prices in hundreds of dollars instead of the military millions. They can be used by anyone who can type plain language on to a standard keyboard, and in any message transmission system on land or in space.

The success of such systems is based on two characteristics: they are inaccessible, in that they do not depend on third parties, such as couriers, to carry secret decoding information between sender and receiver; and they are economically invulnerable, in that they make the interceptor's task so difficult that it becomes beyond his means to undertake the task of decoding. Neither of these essential characteristics was available to the commercial world (nor, for that matter, to the military world) until the advent of digital communications, followed by the ability of microprocessors to manipulate data streams and the application of clever mathematics to that manipulation.

Most of the new systems in use are based on the American Data Encryption Standard (DES) which was developed from mathematical work generated by IBM, MIT and others. They permit the sender to encrypt his message automatically at very high speed, and the intended receiver to decrypt at the same speed, without any secret codes being passed between them, and yet they give the interceptor an impractical task in decryption – even though the encoding or decoding keys may be openly published in a telephone directory!

The core of this apparently impossible feat is derived from two types of mathematical computation, both of which go back to Euclid. The first is known as *one-way functions*, and the second as *modulo arithmetic*.

One-way functions

One-way mathematical functions are those in which it is easy to compute in one direction but very complex to compute in reverse. The best-known example is called the *knapsack problem*, which is best envisaged by thinking about wooden disks that have to be dropped into a cardboard tube.

Imagine a cardboard tube, 10 cm in diameter and 1 metre long, and a lot of wooden disks, just under 10 cm diameter so that they can drop into the tube, and of varying thicknesses. The thinnest might be 1 mm thick, the thickest might be 100 mm thick, and between them is a whole range of thicknesses that might be 2 mm, 4 mm, 7 mm, 10 mm and so on. The sender of the message (each letter in the message will need one tube) drops a certain number of disks of different thickness into the tube, using a simple list of numbers which tells him how many disks of each type he must put in. There may be, for instance, 120 one mm disks, 13 two mm disks, no four mm disks, 32 seven mm disks, and so on. When the correct number of disks have been dropped in, he cuts the top of the tube flush with the last disk, seals both ends with sticky tape, and sends it on its way. Computing in this direction has been very simple.

However, an interceptor has to compute in reverse. He grabs the tube as it flies by, measures its length (because it is the length that determines what coded letter is inside) and lets it continue its journey with its end seals unbroken. It measures, say, 900 mm. Now, what was inside? 900 one mm disks? Or 9 one-hundred mm disks? *Or any combination in between?* What he has to do, even knowing the original list of numbers that the sender used to put in the right quantities of disks, is to try all possible combinations of all the disk thicknesses until he finds a combination that adds up exactly to a length of 900 mm. If the original list of numbers used by the sender was ten numbers long, and there are twenty different thicknesses of disk, he has to undertake 20^{10} calculations, that is, ten million million tries. Even at a rate of a million calculations a second (which would need a very powerful computer), that would take him four months, just to extract one letter from the message.

At the end of its journey, the tube lands on the desk of the intended recipient. He can measure the length and deduce the contents as quickly as the sender put them in, using a list of secret numbers that are known only to him and no one else, not even the sender. He is using modulo arithmetic.

Modulo arithmetic

Modulo arithmetic is a mathematical process by which smooth progressions of numbers can be turned into a series of new numbers that jump about in an apparently random fashion, and then it can convert those erratic numbers back into the smooth series again. It is presented as

$$M = a \text{ modulo } b$$

where M is the 'left-over' after a has been divided by b. For example, if a is 22 and b is 13, a divided by b is 1 with 9 left over, so $M = 9$.

When this process is applied to a smooth series, an apparently random series appears:

a =		100	120	140	160	180	200
b	= 23						
M =		8	5	2	22	19	16

What modulo arithmetic does in encryption is to use this apparent randomness to complicate further (for the interceptor) the process of decrypting one-way functions, while leaving the sender and the receiver with very simple tasks in encryption and decryption.

The combined use of one-way functions and modulo arithmetic is best understood by considering a very simple example.

Example of encryption and decryption

Assume first a 5-bit code for the message. This is a digital form of the alphabet, which gives 2^5, that is, 32 binary codes, which is enough for the

26 letters and a few punctuation marks. Numbers can also be included if they are preceded by a coded 'NUM' meaning 'numerals follow' and ended with 'MUN'. So we can have, for example,

$$\begin{array}{l} 00001 = A \\ 00010 = B \\ 00011 = C \\ 00100 = D \\ 00101 = E \\ 00110 = F \\ \quad\vdots \qquad \vdots \end{array}$$

If a space between words is allocated 00000, the message 'A BAD FACE' would appear as a digital stream:

00001000000001000001001000000000110000010001100101

An interceptor, given this message un-encrypted and a few dozen other English words coded in the same way, could decode the entire alphabet in a few minutes, using classic decoding techniques of letter and word frequencies and general logic.

So let us encrypt just one letter in this example, say 'S', which might be coded 11010 in the 5-bit code. The intention is to disguise 11010 so that it appears as a 'one-way' function to the interceptor, while the process of generating that function must be easy for the sender, and decrypting it must be easy for the receiver, without either of them exchanging any secret information.

The process works as follows. The receiver holds, in his own decryption processor, a series of five numbers, which are his own secret and which he reveals to no one. He also holds two other numbers, which are also his own secret, and which will be used for modulo arithmetic. These last two numbers can be chosen at random, preferably as prime numbers, but the other five, while random, must be such that they ascend in an order in which each number is not less than the sum of the previous two. Thus, his five numbers could be

3 5 9 16 25

while his other two numbers (to be used for a modulo b) are $a = 109$ and $b = 544$. From a and b he calculates and holds an inversion of the modulo, that is, the number which, when multiplied by a modulo b, produces 1. This number is 5. ($5 \times 109 = 545$; 545 modulo $544 = 1$.)

The receiver now calculates his reception code, which will be

$$\begin{aligned} 3 \times 109 \text{ modulo } 544 &= 327 \\ 5 \times 109 \text{ modulo } 544 &= 1 \\ 9 \times 109 \text{ modulo } 544 &= 437 \\ 16 \times 109 \text{ modulo } 544 &= 112 \\ 25 \times 109 \text{ modulo } 544 &= 5 \end{aligned}$$

and he openly publishes the reception code as 327, 1, 437, 112, 5.

Anyone sending to that receiver now knows that, for each 5-bit block that he sends, he must multiply the first bit (1 or 0) by 327, the second bit by 1, the third by 437, and so on. So our letter 'S', coded 11010, becomes

$$\begin{aligned} 1 \times 327 &= 327 \\ 1 \times 1 &= 1 \\ 0 \times 437 &= 0 \\ 1 \times 112 &= 112 \\ 0 \times 5 &= 0 \\ & \overline{440} \end{aligned}$$

440 is the length of the equivalent cardboard tube seen earlier. In fact, of course, it is transmitted in binary form using the 5-bit code for numbers.

When the interceptor receives the 440 (he has already broken the 5-bit code so he knows the digits for '4', '4' and '0') he has to undertake 2^5, that is 32 calculations, multiplying each published code number by 1 or 0 until he gets the right addition, thus:

$0 \times 327 = 0$		$1 \times 327 = 327$		$1 \times 327 = 327$
$0 \times 1 = 0$		$0 \times 1 = 0$		$1 \times 1 = 1$
$0 \times 437 = 0$	. . .	$1 \times 437 = 437$	. . .	$0 \times 437 = 0$
$0 \times 112 = 0$		$0 \times 112 = 0$		$1 \times 112 = 112$
$1 \times 5 = 5$		$1 \times 5 = 5$		$0 \times 5 = 0$
total 5		769		440
NO		NO		YES

On average, he would need to do only 16 such calculations, because he might stumble on the right answer at the first try ('A') or have to go right through the 32 combinations.

Remember that, in real life, the interceptor's task is immense. He is not faced with additions of five numbers but perhaps of 50, adding up to totals such as 18 763. With bit strings 50 bits long, instead of the 5-bit string in this example, he is faced with 2^{50} possible additions, that is an average of 500 million million calculations per message letter. At the million calculations per second that he can manage with his big computer, each letter would take him about 20 years.

Now we turn to the intended recipient, who has also received the 440, and who now puts modulo arithmetic into play, using his secret numbers which are 3, 5, 9, 16 and 25, his second modulo number b, which is 544, and his inverse modulo, which is 5.

Firstly, he takes his inverse modulo 5 and uses it to form a new modulo:

$$440 \times 5 \text{ modulo } 544 = 24$$

In this way, he has now converted the 440 from its random multipliers to a much smaller number to which he can apply his small secret numbers. But it is not the relative reduction in the size of the total that saves him time, because there is the essential mathematical trick yet to come.

Remember that his five secret numbers had to be in a series in which each number is not less than the sum of the previous two. Because of this, he does not have to go through even the 32 calculations that the interceptor suffered – he manages with just five, using simple logic steps. He is looking for a total of 24, made up from bits that are valued 3, 5, 9, 16 and 25 in a 5-bit code. Therefore:

1 The 5th digit must be 0, because its value of 25 exceeds the total of 24.
2 The 4th digit cannot be 0, because the remaining values (3, 5 and 9) do not make up the total. Therefore it must be 1, with a value of 16, leaving 8 to find.
3 The 3rd digit must be 0 because its value exceeds 8.
4 The 2nd digit cannot be 0 because the only remaining digit (the first) does not make up the total. Therefore it must be 1, with a value of 5, leaving 3 to find.
5 The first digit must be 1, with its value of 3.

Therefore the letter is 11010, which is 'S'.

That is a very simple example of how modulo arithmetic can turn a simple one-way function (which can be solved by logic steps) into a complex one-way function, which can be solved only by total analytical treatment, and then turn it back again *without revealing its secret numbers.*

Note the difference in tasks in real life, as distinct from this example. With block totals made up of 50-bit strings of digits, the interceptor has to make 2^{50}, that is 500 million million calculations. The receiver has only to do 50. Also, in reality, the numbers used are much more complex, which makes the interceptor's task correspondingly harder, but not that of the sender or receiver.

Note too that classic code-breaking techniques, such as seeking the location of most-frequent letters, cannot be applied. For instance, 'E' is the most frequently used letter in the English language. But if the phrase 'THE END' is digitised and sent as a 35-bit string in a 5-bit code (6 letters plus a space), the first E takes digits 11 to 15, while the second E takes 21 to 25. Each of those digits will be subject to a different multiplier, so the total values of the two Es will be different, and both will be buried in a much larger total.

Thus the apparently impossible feat has been accomplished: easy encryption, easy decryption; an interceptor's task that grows exponentially with the number of bits sent in each data block; no revealing of secret codes – indeed, the open publishing of one side of the code; and all for the price of a typical word processor at each end of the circuit. This is why potential users of commercial space communications need not be concerned about message security in the future.

Scrambling

Digital encryption can be applied to all forms of digital transmission, which now encompass voice, data and video used for teleconferencing. High-quality TV for entertainment purposes is not yet digitised – it is still transmitted by frequency modulation of the carrier wave – so digital encryption cannot yet be applied. To make TV transmissions secure, for such applications as pay-TV, the current practice is to 'scramble' the picture by disturbing the line-synchronisation so that the picture breaks up and cannot be reassembled without an appropriate descrambler coupled into the TV set. This form of scrambling can be overcome relatively easily by a dedicated electronic eavesdropper, but there is increasing use of digital encryption techniques applied to the processes that disturb line-synchronisation and to the sound associated with the picture. Together, these present a much more difficult problem to the 'hacker', who will find his decoding problems magnified even more when digital TV arrives.

15

Operational systems

This chapter shows a selection of commercial satellite systems that are currently in operation or that are planned to become operative during the late 1980s. Each map shows the EIRP contours delivered by the satellite so that, using the examples presented in Chapter 11, Earth station sizes for different qualities and bandwidths of traffic can be calculated, or qualities and bandwidths deduced from given antenna sizes. Satellite G/T is also shown, so that, again using Chapter 11, up-link calculations can be made. Satellite longitude, in degrees East or West of the Greenwich meridian, is also indicated, so that, if the latitude and longitude of an Earth station is known, its pointing angles in elevation and azimuth can be read from Figure 61.

The EIRP values (all in dBW) and contours are approximate but are adequate for the purpose of estimating the quality of service that may be expected within a given coverage area. Satellite G/T values are also approximate.

Further, more detailed information may be obtained from the satellite owners, whose names and addresses are given in the appendix.

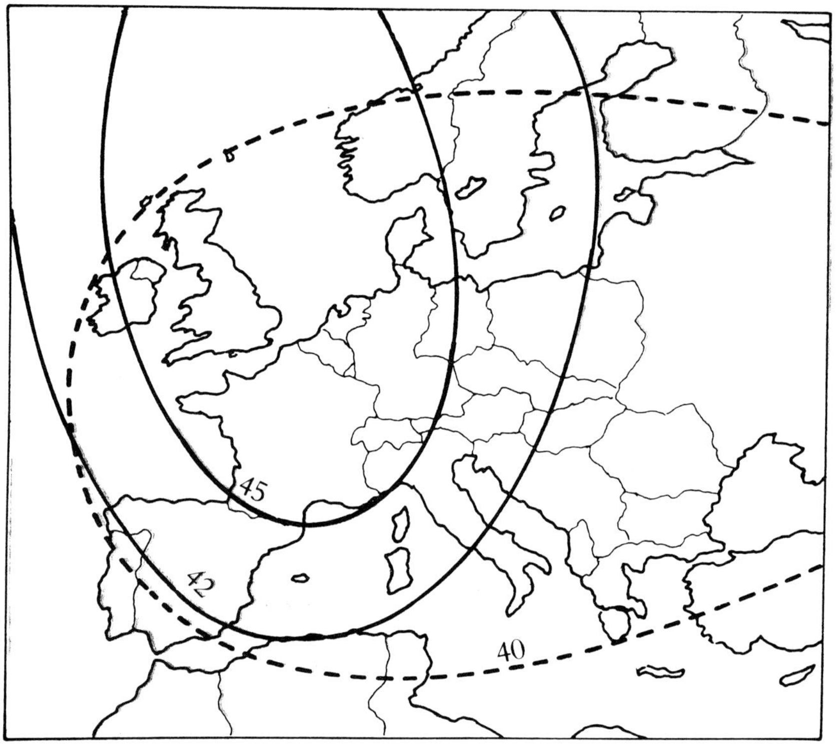

Figure 64 *Satellite name:* ECS
Owner: Eutelsat
Satellite locations: 10°E 13°E
Satellite G/T: −3 dB/K *solid lines*
Frequencies: −0.5 dB/K *dotted line*
Solid lines 14 GHz up 11 GHz down
Dotted line 14 GHz up 12 GHz down
Service start date: 1983

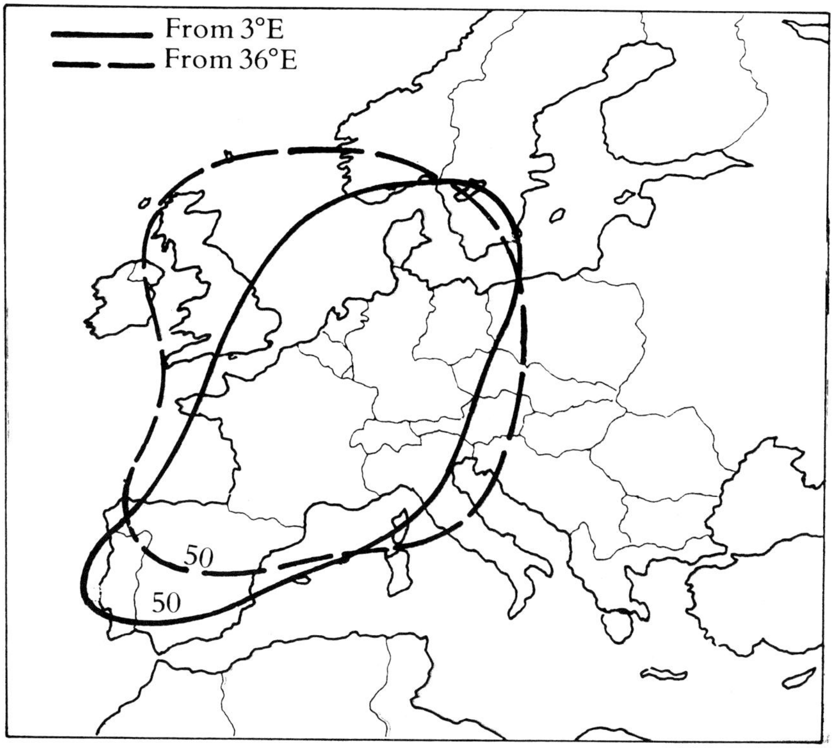

Figure 65	*Satellite name:*	*Eutelsat*-2	
	Owner:	Eutelsat	
	Satellite locations:	3°E 36°E	
	Satellite G/T:	2.5 dB/K	
	Frequencies:	14 GHz up	11 GHz down
	Service start date:	1984	

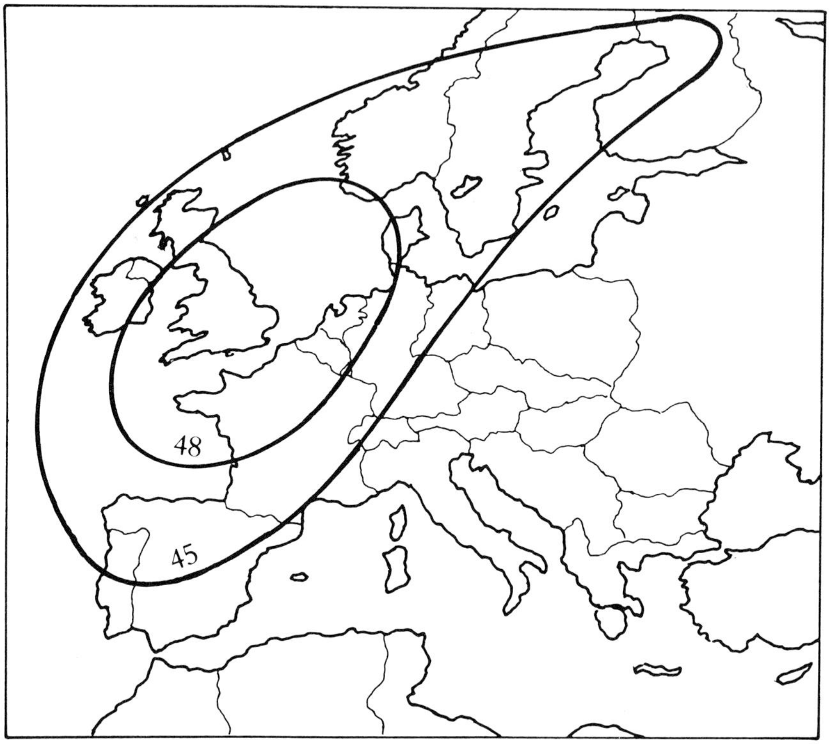

Figure 66	*Satellite name:*	*Intelsat* V, F10
	Owner:	Intelsat
	Satellite location:	25°W
	Satellite G/T:	0 dB/K
	Frequencies:	6 GHz up 4 GHz down
	Service start date:	1985 (1980 for Intelsat V series)

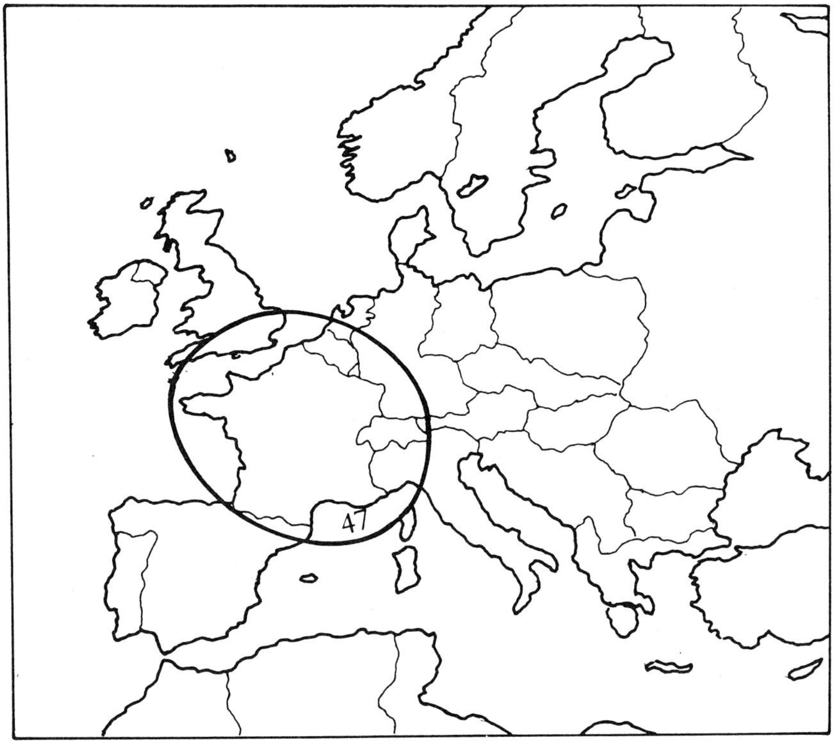

Figure 67 *Satellite name:* *Telecom* 1
Owner: DGT, France
Satellite locations: 5°W 8°W
Satellite G/T: 6.5 dB/K
Frequencies: 14 GHz up 12 GHz down
Service start date: 1984

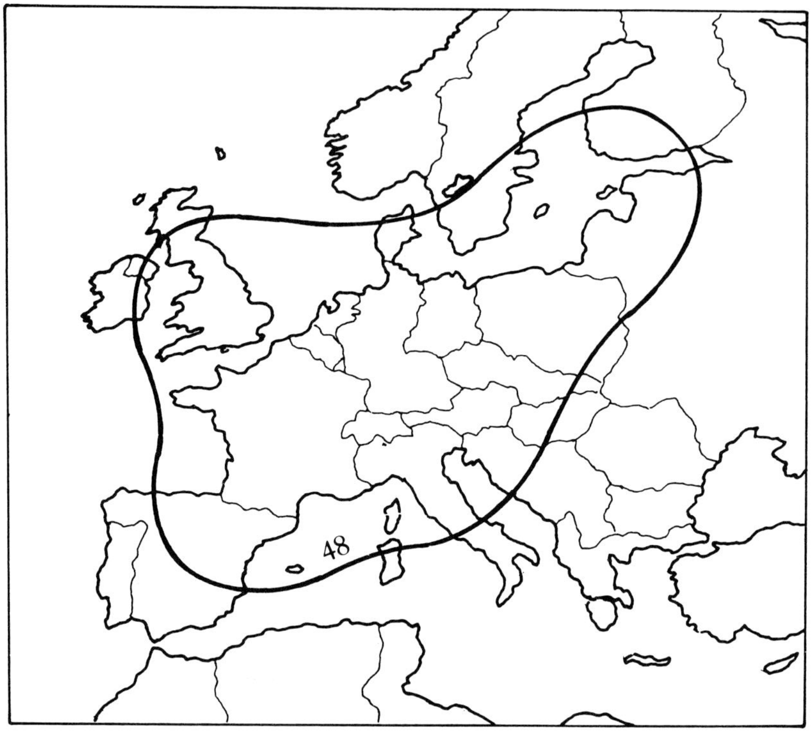

Figure 68

Satellite name:	*Astra* 1
Owner:	SES, Luxembourg
Satellite location:	19°E
Satellite G/T:	NA
Frequencies:	14 GHz up 11 GHz down
Service start date:	1988 planned

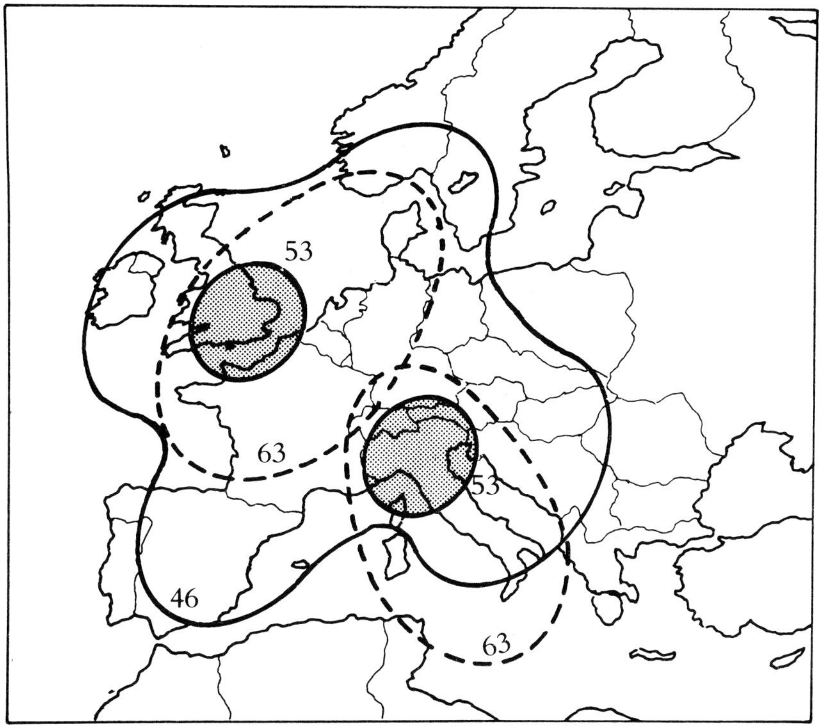

Figure 69 *Satellite name:* *Olympus* 1
Owner: European Space Agency
Satellite location: 19°W
Satellite G/T: 14 GHz 5.6 dB/K
18 GHz −0.6 dB/K
30 GHz 10.9 dB/K
Frequencies: *Solid line* 14 GHz up 12 GHz down
Dotted lines 18 GHz up 12 GHz down
(Direct broadcast TV)
Shaded beams 30 GHz up 20 GHz down
All beams are steerable
Service start date: 1988 planned

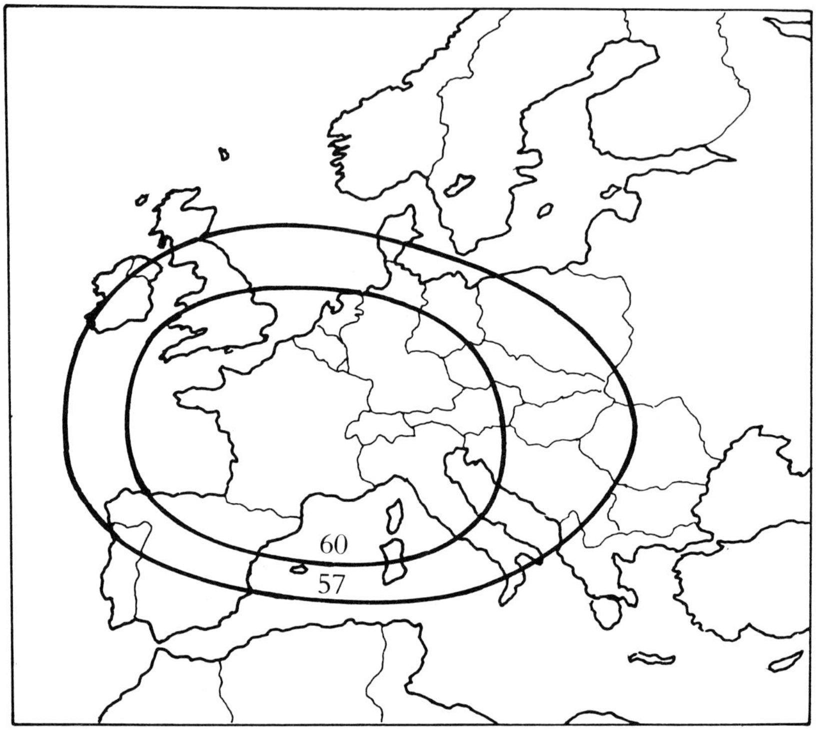

Figure 70	*Satellite name:*	TDF 1
	Owner:	TDF, France
	Satellite location:	19°W
	Satellite G/T:	NA
	Frequencies:	18 GHz up 12 GHz down (Direct broadcast TV)
	Service start date:	1987

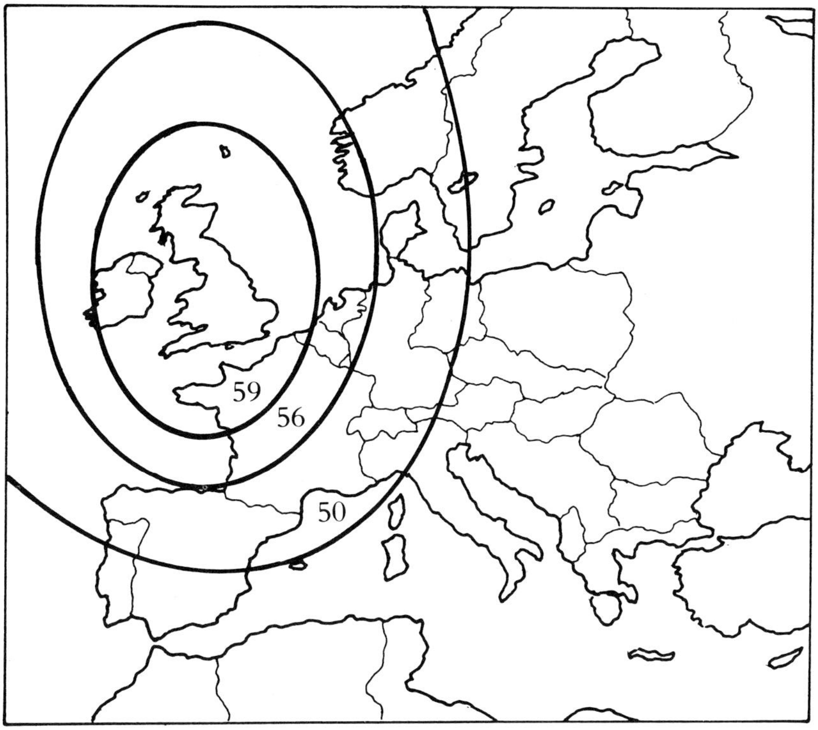

Figure 71	*Satellite name:*	To be named
	Owner:	BSB (UK)
	Satellite location:	31°W
	Satellite G/T:	NA
	Frequencies:	18 GHz up 12 GHz down (Direct broadcast TV)
	Service start date:	1990 planned

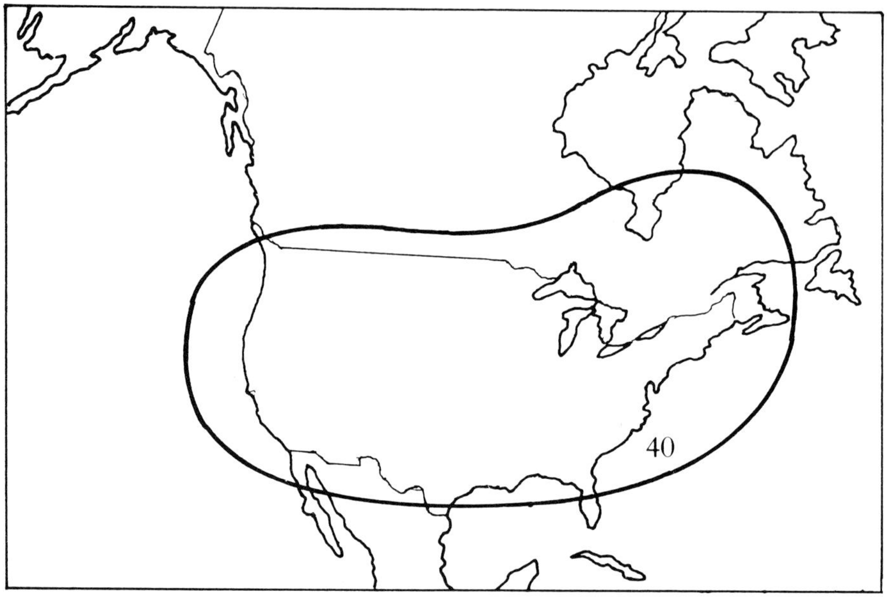

Figure 72	*Satellite name:*	SBS
	Owner:	Satellite Business Systems, USA
	Satellite locations:	95°W 97°W 99°W 101°W
	Satellite G/T:	2 dB/K
	Frequencies:	14 GHz up 12 GHz down
	Service start date:	1981

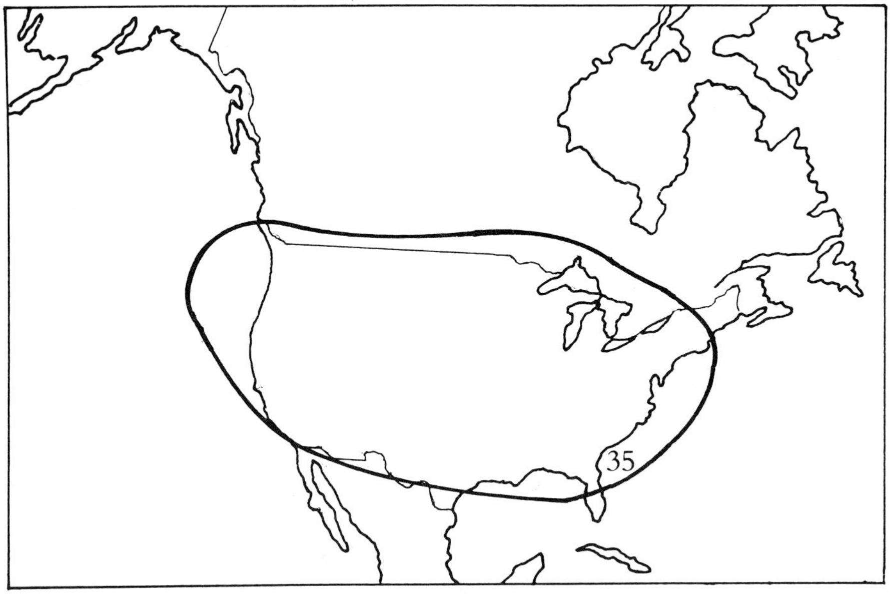

Figure 73	*Satellite name:*	GTE *Spacenet*
	Owner:	GTE Spacenet Corporation, USA
	Satellite locations:	69°W 120°W
	Satellite G/T:	−4 dB/K
	Frequencies:	6 GHz up 4 GHz down
	Service start date:	1984

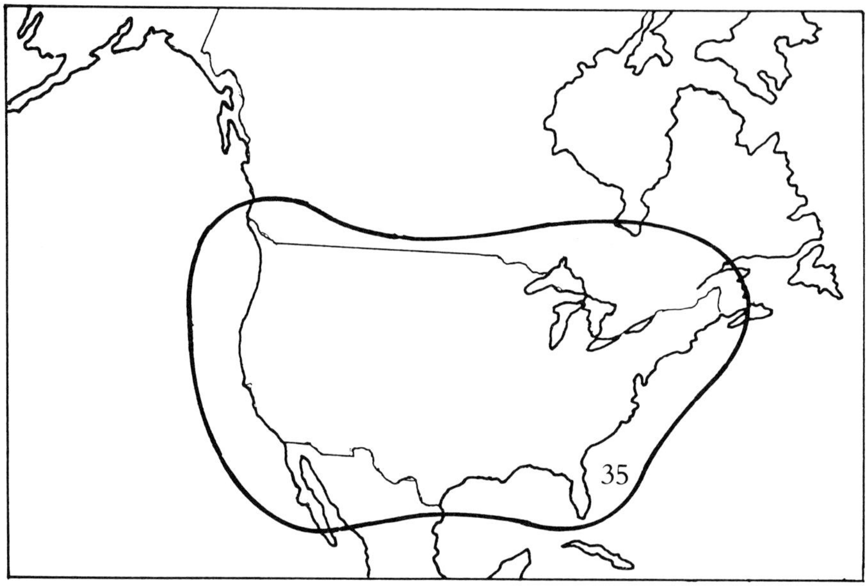

Figure 74	*Satellite name:*	*G-Star*
	Owner:	GTE Satellite Corporation, USA
	Satellite location:	103°W
	Satellite G/T:	−1.5 dB/K
	Frequencies:	14 GHz up 12 GHz down
	Service start date:	1985

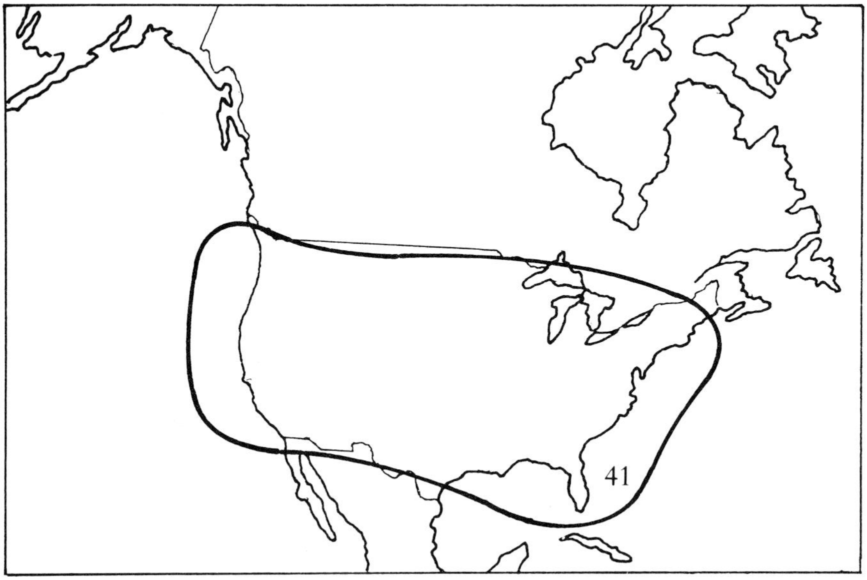

Figure 75	*Satellite name:*	*Satcom Ku*
	Owner:	RCA Americom, USA
	Satellite location:	85°W
	Satellite G/T:	0 dB/K
	Frequencies:	14 GHz up 12 GHz down
	Service start date:	1985

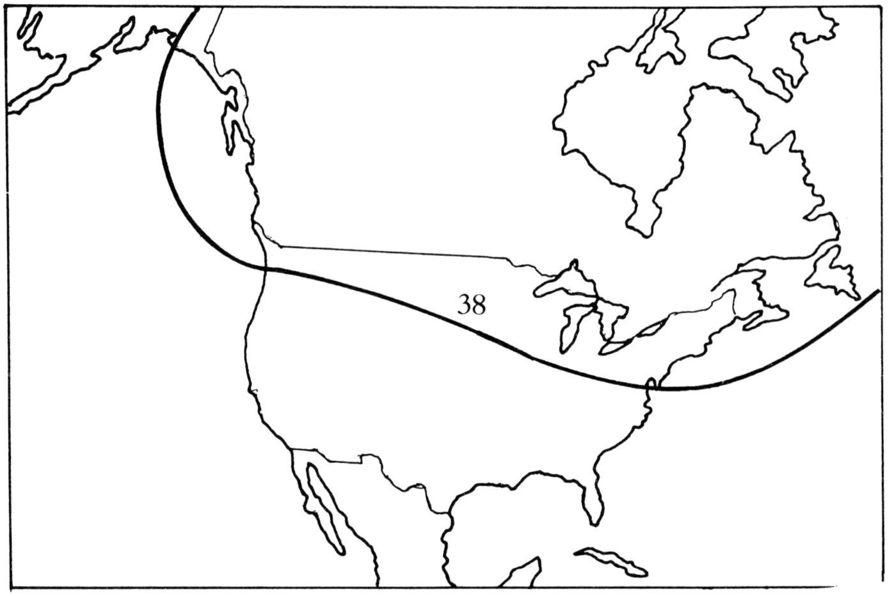

Figure 76	*Satellite name:*	*Anik D*
	Owner:	Telesat, Canada
	Satellite locations:	105°W 112°W
	Satellite G/T:	0 dB/K
	Frequencies:	6 GHz up 4 GHz down
	Service start date:	1982

Figure 78 (*facing*)	*Satellite name:*	*Palapa*
	Owner:	Perumtel, Indonesia
	Satellite location:	108°E
	Satellite G/T:	−7 dB/K
	Frequencies:	6 GHz up 4 GHz down
	Service start date:	1983

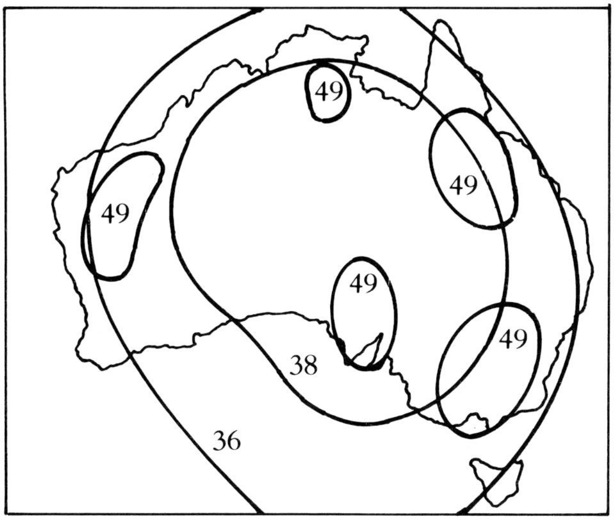

Figure 77 *Satellite name:* *Aussat* 1
Owner: Aussat, Australia
Satellite locations: 156°E 160°E
Satellite G/T: −3 dB/K
Frequencies: 14 GHz up 12 GHz down
Service start date: 1985

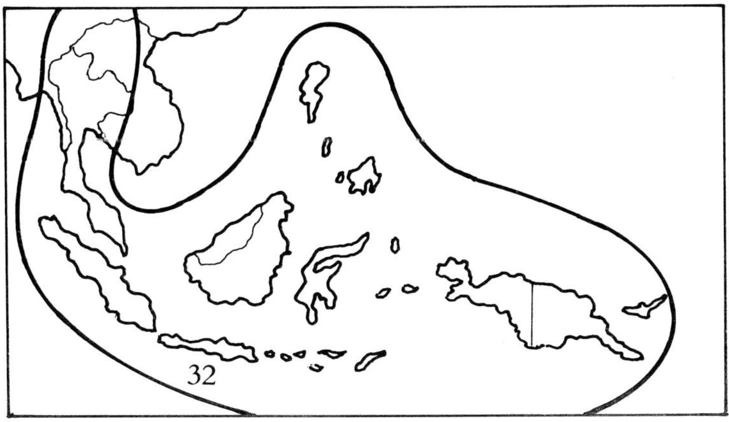

16

The economics of satellite communications

If communications satellites could not be run profitably, no one would buy them. But they certainly are bought, to the extent that, at the end of 1986, there were 80 commercial communications satellites operating in geosynchronous orbit, making profits for their owners, with the population heading towards 100 in the next few years.

To those outside the space industry it comes as something of a shock to learn that, at late-1980s prices, those satellites cost around \$60 million each; that it will cost another \$40 million to launch each one; and that insuring them against failure – particularly after the series of launcher disasters in 1986 – may add another 20% to the bill, so that the final price of the satellite in orbit lies around \$120 million.* How, they ask, can you make profits on that?

The aim of this chapter is to show how.

Satellite prices

Custom-built products are notoriously expensive, and satellites are custom-built. There are no production lines in the satellite industry – the longest production line of identical satellites that can be expected by a manufacturer is a line of three: two in orbit (one acting as standby for the other), and one kept on the ground as a temporary spare in case one of the first two is dropped in the sea by a launcher failure. Large international

*In this chapter, all costs and prices are shown in US dollars, because this is the reference currency used by manufacturers and users of space communications.

customers, such as Intelsat, do order longer runs, perhaps up to a dozen. However, since each of these must be capable of working from a variety of positions in orbit, they require correspondingly complex antennas, and beam-shaping and internal switching facilities; thus the actual cost may be greater rather than less.

Of course, manufacturers take advantage of the 'commonality' of many of the subsystems of their spacecraft – structures, propulsion, power control systems, and so on – to produce standard parts. But even so they must, for each customer, design and manufacture specific antennas for his required coverage, solar arrays to produce the required power, and the entire communications payload, which may vary from a few dozen low-power transponders for telecommunications to a few very-high-power transponders for direct broadcasting of television. And with each change comes the need to verify mass and power relationships, battery capacity, thermal control in orbit, orbital mechanics, and tracking, telemetry and command equipment to match the customer's present and future facilities on the ground.

For this, the manufacturer needs to employ engineers of all disciplines – electronic, electrical, mechanical, structural – designers, physicists, mathematicians, and a whole range of supporting staff. He must have manufacturing facilities that range from super-clean assembly rooms to vast integration halls, test equipment that can vibrate and load an entire spacecraft, heat it, subject it to high vacuum and flood it with simulated solar radiation, and outdoor ranges that can test the beam patterns of new antennas.

From the time a manufacturer receives an order for three spacecraft, it takes about six months for the engineers and designers to produce the drawings for the new parts and subsystems that will be required by the new satellite. While this is going on, the standard parts are beginning to be assembled. Manufacture of the new parts and systems will take about twelve months before they arrive for integration into the parts already built. That is 18 months gone. System testing and full integration will take a further nine months, followed by three months of total testing of the entire satellite. Thirty months have passed. The next three months are spent on further testing of the spacecraft in conjunction with all its ground control equipment that will travel with it to the launch site, and with long-distance connections to the actual ground stations that will control it as it goes into orbit. Finally, at 33 months, the satellite and all its ground support equipment is crated up for air shipment to the launch site, where it is tested again, first on its own and then again when it is stowed in the launcher. From order to launch has taken 36 months.

Over those three years, the manufacturer has employed a team that has probably peaked at 500 people, and for each spacecraft he has spent around $25 million on raw materials, subcontractors and the running costs of his manufacturing and test facilities. That is why, allowing for a little profit, he charges $60 million.

System prices

Except in the United States, communications satellites may be owned only by national telecommunications authorities, or groups of those authorities acting through an international agency, such as Intelsat, Eutelsat and Inmarsat. In the United States, anyone may own a satellite system, but ownership is usually confined to those who can not only raise the capital but also offer the supporting communications infrastructure on the ground. Thus, in the United States, one finds owners such as Western Union, RCA, Comsat and Hughes providing public voice, data and TV across the continent, but new systems are coming that will be dedicated to the provision of such services to commercial customers through very small Earth stations located on their premises. Let us examine the price of such a system.

The satellites

These are of medium size, such as *Eurostar* (Chapter 12) and each will carry twenty-four 40-watt transponders to provide an EIRP of 45 dBW over a designated coverage area (Chapter 9). They will cost $60 million each. Two satellites are required in orbit; the second acts as standby in the unlikely event of total failure of the first, but it can be used for traffic until that possible failure occurs. In this case its capacity is offered on what is called a 'pre-emptible' basis, that is, at a cheaper rate than normal, because it may be withdrawn from service to take over the traffic of the failed prime spacecraft. Since satellites take three years to build, it is usual to order a third, part-built as a spare, to be launched as soon as possible if one of the first two suffers a launch failure. That is two-and-a-half satellites, totalling $160 million.

Launchers

Eurostar, like all current communications satellites, is designed to be launched by either the European Ariane rocket or the US Shuttle. In either case, launching each satellite will cost around $40 million, that is a total of $80 million for the two into orbit.

Insurance

The world of insurance took an optimistic view when it began to underwrite insurance of spacecraft against launch failures, and early premiums in the late 1970s stood at around 7% of the launched cost (that is, the cost of the satellite plus the launcher). Since then, after a series of launch disasters that peaked in 1986, insurance premiums have risen steeply and the best that can be obtained today is about 20% of the launched cost. Pre-flight insurance for two spacecraft and two launchers with a total value of $200 million will therefore cost a further $40 million.

Summarising so far, we have to find:

		$ million
Satellites 60 + 60 + 40	=	160
Launchers 40 + 40	=	80
Insurance 20 + 20	=	40
		280

That total 'up-front' money of $280 million has to be obtained from one or more of three sources: internal funds, equity investment, and bank loans. In each case the going rate of money will apply, and over a 10-year period dividends to investors and interest to the banks will amount to around $150 million.

Finally, the owner has to take responsibility for keeping his spacecraft on station at their allocated orbital locations, and for this he will need a tracking, telemetry and command (TTC) centre that will require land, at least two large antennas, computers, and all the engineering and operations staff necessary to monitor and control the two spacecraft 24 hours a day. This, together with the management, administration and marketing staff that are needed, will cost the owner, over 10 years, another $50 million.

So the total system cost to the owner over ten years will be:

	\$ million
Launched and insured spacecraft	280
Cost of capital	150
TTC station and management	50
	480

Taking a simple average, this works out at \$48 million per year.

System income and profit

Why is it worth spending \$48 million per annum on a satellite system?

Ignoring for the moment the standby satellite, we have 24 active transponders in orbit. The going 'wholesale' price for a transponder in the USA today is around \$3.5 million p.a., so we break even if we sell 14 of the 24. Selling 20 brings in a profit of \$22 million per year; filling the satellite brings an annual profit of \$36 million. Currently USA satellites operating in C-band and Ku-band are between 85% and 90% full.

If the standby satellite transponders are offered at half price, and 50% of them are filled, there is an additional non-guaranteed profit of \$21 million per year.

Of course, account has to be taken of adverse cash flows in the early years while the satellites are being filled by a growing number of customers, but it is clear that, to the investor and the banker, the total profit potential is very attractive indeed.

User profit

The annual transponder price of \$3.5 million per year represents about \$500 per hour. For that price, a television company can send an hour's programming to every cable operator in the coverage area, which will encompass many millions of homes. The price by terrestrial links can be hundreds of times higher.

User savings are equally impressive in the world of corporate communications. As an example, take a typical American company with 500 sites spread across the USA that are currently linked by terrestrial means. The average cost of a long-distance telephone call charged to that company, via the local telephone operating companies near each site and via the AT&T 'long lines', is 45 cents per minute at mid-1980s prices. Let us calculate the savings that would accrue if that company installed a small Earth station at each site and hired a transponder for its communications, so bypassing the terrestrial circuits:

1 The retail hire charge for the transponder from the satellite owner will be $4 million per annum.
2 The cost of each Earth station will be $10 000 installed, totalling $5 million, and the user will need a large 'hub' station to control the network and to provide good overall *C/N*. This will cost another $0.5 million, so the total capital cost of the ground equipment is $5 500 000, which is an annual cost over ten years of $550 000.
3 The cost of borrowing that capital over 10 years will be about $3.5 million, which equates to $350 000 per annum.

Thus, the total annual cost to the user will be

$$\$4\,000\,000 + 550\,000 + 350\,000 = \$4\,900\,000$$

which we can call $5 million, or $10 000 per site per year.

As seen in Chapter 11, a 36 MHz transponder can handle 54 Mb/s using QPSK modulation. Thus, if every station in the system shared equally the transponder all of the time, each could take 100 kb/s for 24 hours per day. This means three voice channels at 32 kb/s, or 10 data channels at 9.6 kb/s, or one video-conference channel at 56 kb/s. Taking the voice channels for comparison with the terrestrial circuits, the three channels will cost, for each site, $10 000 per year, which is $40 per working day. With an 8-hour day, that is $5 per hour, which is 8 cents per minute. A two-way voice conversation needs two channels, so the final cost of a two-way circuit is 16 cents per minute. This can be compared with the terrestrial cost of 45 cents per minute.

In practice, of course, each station would not be sending and receiving identical traffic patterns all of the time. The satellite system provides the flexibility that permits one station to take, say, 2 Mb/s for a specific time, while the other stations are not using their available capacity, and this is an added advantage that cannot be offered by terrestrial systems, in which the bandwidth laid to each site is effectively fixed.

The practice of bypass, with the level of savings that can be seen, is being increasingly adopted in the United States, where hundreds of companies are installing thousands of small Earth stations for this purpose.

What is not so well recognised is the economic effect of the immediacy of space communications. Immediacy means having the ability to make wideband connections between two or more sites at once, by driving transportable stations on to those sites, instead of having to wait months, or even years, for cables of equivalent bandwidth to be laid. An example was seen recently in a large corporation, with many sites, which installed a large new computer facility at one site to handle computer-aided design and manufacture (CADAM). Knowing that the capacity of the system would not be filled for a year or two, the site offered its facilities to others until they subsequently installed their own CADAMs. One of the sites investigated the offer and found that it would be at least two years before the necessary 2 Mb/s cable could be laid between the two locations. So the corporation installed two temporary Earth stations, and the second site started using CADAM immediately.

The interesting point about this story is that the two sites were only 15 miles apart, which belies the oft-stated doctrine that 'the minimum economic distance for satellite communications is 2000 miles'.

17

Future trends

This book has only skimmed the surface of some of the principal technologies employed in space communications systems. It has attempted to show how those technologies have brought satellites to the stage of being essential components in the expanding world of television and telecommunications, and how the space system designer can use them and exploit them to the full.

The technologies will continue to improve the performance of space systems, and of terrestrial systems as well, and both systems will be seen working together into the twenty-first century. Some of the trends that can be foreseen today are briefly reviewed here.

Satellite size

Until the advent of geosynchronous space stations towards the end of the century, it is unlikely that the geosynchronous orbit will see communications satellites larger than the biggest one under construction today – the British Aeropace *Olympus*. Over the next fifteen years or so, such satellites will continue to be launched by Ariane and by Shuttle, both of which impose constraints on physical size and lift-off mass. However, within those constraints, technological improvements can be expected which will improve satellite performance, as the following sections indicate.

Sun-pointing satellites

Designs already exist for sun-pointing spacecraft in which the body, instead of being 'locked-on' to Earth, as with a conventional three-axis satellite, is pointed permanently at the Sun, while the antennas, which must keep looking at the Earth, rotate once per day on the body. The advantage lies in the fact that the body is permanently shaded from the sun (except for one face) by its own solar arrays, and thus all the body faces can be used to radiate heat instead of just the two north–south faces as in a conventional satellite. In comparison with a conventional body of the same size, a sun-locked body can radiate twice as much heat, and because of this it can accept twice as much power from its arrays and radiate twice as much RF power.

Solar arrays

The conversion efficiency of solar cells has steadily increased over the past twenty years, and the increase can be expected to continue, particularly with the use of new materials, such as gallium arsenide, for the cell materials. Arrays of the 1990s can be expected to produce 20% more power than arrays of the same size today.

Batteries

Ever since the construction of the first satellites in the late 1950s and early 1960s, batteries have been a plague to the satellite designer. Today, the efficiency and reliability of spacecraft batteries are significantly higher than those of the early days, and their mass–power ratio is improving all the time. These improvements, particularly seen in the pressurised metal–hydrogen batteries now coming into use, can be expected to continue, saving mass that can be more usefully exploited in the communications payload.

Transponder efficiency

Most TWTs today exhibit DC to RF conversion efficiencies of less than 40%. Solid-state amplifiers are more reliable than TWTs, but their conversion efficiency is even lower, and they are more susceptible to temperature changes than are TWTs. The world's manufacturers of travelling wave tubes and solid-state amplifiers are expending continuous effort to improve efficiencies, which will permit the satellites to radiate more RF watts per watt of DC from the solar arrays, but progress is slow. Other components in the transponders, particularly filters and multiplexers (which are arrays of filters) are notably inefficient, also radiating over 60% of their input power as waste heat. As with amplifiers, no significant improvements can be foreseen.

Antennas

Satellite antennas will get larger as smaller beams are called for to provide higher EIRPs and to generate complex beam shapes. Typical diameters of 1 to 3 m today will grow to 5 to 15 m in the 1990s, using unfolding and erecting mechanisms that are now in development. In the low-frequency bands allocated to mobile communications – that is, to ships, aircraft and vehicles – much larger sizes will be seen, perhaps beyond 50 m diameter. These are likely to be constructed by new non-mechanical techniques, such as balloon antennas, erected by gas when in orbit and then stiffened by the injection of rigidising foam or by self-stiffening under ultra-violet radiation from the Sun. Beam steering is already practised by steering complete antenna assemblies. This will be complemented by the increasing use of multi-feed phased arrays, which will allow remote beam shaping as well as steering, and the ability to 'hop' beams from one point in a service area to another, at millisecond rates, according to traffic demands.

Satellite switching

SS-TDMA (Chapter 13) will become common practice, with the satellite becoming a veritable 'telephone exchange in the sky', reducing the complexity, and thus the cost, of TDMA Earth stations.

Operating frequencies

Ka-band (18 to 30 GHz) will be seen in increasing use in commercial communications systems over the next decade, bringing the advantages of smaller antennas, both on the satellite and the Earth stations, for given gains, and freedom from interference from and with terrestrial radio systems.

Low noise amplifiers

The noise performance of low noise amplifiers, both in Earth stations and in satellites, will continue to improve, although the rate of improvement, which has been substantial since the 1970s, must slow down as the amplifiers approach their theoretical minimum noise performance.

Modulation and compression techniques

Digital modulation has permitted reductions in bandwidth to proceed at a phenomenal pace, and further reductions can be expected, without sacrificing quality, until the theoretical minima are achieved for each class of circuit.

Fibre optics

It is generally accepted that the simplicity, cheapness and bandwidth capacity of optical fibres will lead to their taking the major part of future traffic on trunk digital routes between city centres. However, optical fibres are not economically suited to the wide dispersion of circuits outside city centres, and this requirement for disperson applies to all television broadcasting and distribution and, equally importantly, to corporate telecommunications networks. It is in these applications that satellites will be seen as the economic solution. They will even be seen to offer economic advantages in general telecommunications, because of their ability to bypass the sections of circuits that generate most of the circuit costs – those that lie in local switching and distribution. Even if the costs of fibre optic city centre trunks fell to zero, those circuit end-costs would still apply – but not via satellite.

Economics

Space communications are already recognised as the economic way to broadcast and distribute television, and in the United States the same recognition has applied to corporate telecommunications. This recognition will undoubtedly spread across the Western world, marked by an ever-increasing population of small Earth terminals on factory roofs, office car parks, docks, airports, hotels and homes. They will also be seen in increasing numbers on ships (several thousand already carry them), aircraft, trucks and, in the not too distant future, cars.

The growth is driven by economics – satellites provide cheaper ways of communicating, and cheapness means lower operational costs, which in turn means more profit for the users – the fundamental driving force of business.

Communications satellites are no longer technological toys. They are now simply active components in the world's daily communications traffic in entertainment, commerce and industry. They will become increasingly useful in those fields, with widening recognition of their economic advantages, even though the majority of users will never know much about their technologies. And that is how it should be.

Further reading

Pritchard & Sciulli: *Satellite Communication Systems Engineering*, Prentice-Hall (1986).

A comprehensive textbook that covers all aspects of the subject in great detail.

Bleazard: *Introducing Satellite Communications*, National Computing Centre Ltd (1985).

This deals particularly with data transmission and business services via satellite.

Rainger, Gregory, Harvey & Jenning: *Satellite Broadcasting*, John Wiley (1985).

Broad and detailed coverage of television engineering and TV broadcasting from space.

Maral & Bousquet: *Satellite Communications Systems*, John Wiley (1986).

A textbook on overall system and subsystem design.

Martin: *Communications Satellite Systems*, Prentice-Hall (1978).

The first comprehensive textbook, covering systems of the 1970s.

Taub & Schilling: *Principles of Communications Systems*, McGraw-Hill (1971).

A standard textbook for engineering undergraduates, with emphasis on modulation techniques, information theory and noise.

Appendix
Manufacturers and owners

The Public Relations and Marketing Departments of the manufacturers and owners shown in this appendix will generally provide detailed information on their products and services on request.

Further information on all types of spacecraft and space systems may be obtained from the Public Relations Departments of:

NASA 600 Independence Avenue SW
Washington, DC 20546
USA

ESA European Space Agency
rue Mario-Nikis
75015 Paris
France

Principal manufacturers of spacecraft in Europe

AEROSPATIALE SA	Division Systèmes Balistiques et Spatiaux BP 96 F-78133 Les Mureaux Cedex France
BRITISH AEROSPACE PLC	Space and Communications Division Argyle Way Stevenage, Herts, SG1 2AS England

DORNIER SYSTEM GmbH	PO Box 1360 D-7990 Friedrichshafen 1 Federal Republic of Germany
EUROSATELLITE GmbH	Steinsdorfstrasse 13 D-8000 München 22 Federal Republic of Germany
MATRA SA	Space Branch 10 Avenue Kleber 75116 Paris France
MBB/ERNO GmbH	Huenefeldstrasse 1–5 PO Box 10 59 09 Bremen 1 Federal Republic of Germany
SAAB SPACE AB	PO Box 13045 S-402 51 Goteborg Sweden
SELENIA SPAZIO SpA	Via di San Alessandro 28–30 1-00131 Rome Italy

Principal manufacturers of spacecraft in the USA

FORD AEROSPACE CORPORATION	Western Development Laboratories Div. 3939 Fabian Way Palo Alto, California 94303
GENERAL ELECTRIC COMPANY	Valley Forge Space Center PO Box 8555 Philadelphia, Pennsylvania 19101
HUGHES AIRCRAFT CO.	Space and Communications Group PO Box 92919 Los Angeles California 90009
RCA ASTRO-ELECTRONICS	PO Box 800 Princeton, New Jersey 08540

TRW	Space and Technology Group One Space Park Redondo Beach California 90278

Principal manufacturers of Earth stations in Europe

ALCATEL THOMSON ESPACE	11 Avenue Dubonnet 92407 Courbevoie France
BELL TELEPHONE MANUFACTURING CO. NV	Defence and Aerospace Group Francis Wellesplein 1 2018 Antwerp Belgium
BRITISH AEROSPACE PLC	Space and Communications Div. Argyle Way Stevenage, Herts SG1 2AS England
CIT-ALCATEL	Transmission Department BP 344 22304 Lannion France
FERRANTI ELECTRONICS LTD	Microwave Division First Avenue Poynton Industrial Estate Stockport, Cheshire SK12 1NE England
GTE TELECOMMUNICAZIONI SpA	Viale Europa 46 20093 Cologno Monzese Milan Italy
MARCONI DEFENCE SYSTEMS LTD	The Grove Warren Lane Stanmore, Middlesex H47 4LY England

MATRA SA	Space Branch 10 Avenue Kleber 75116 Paris France
MBB/ERNO GmbH	Space Systems Group Huenefeldstrasse 1–5 PO Box 10 59 09 Bremen 1 Federal Republic of Germany
SIEMENS AG	Hofmannstrasse 51 PO Box 70 00 73 8000 München 70 Federal Republic of Germany
TELSPACE	10 Avenue de Centaure 95800 Cergy St Christophe France

Principal manufacturers of Earth stations in the USA

ANDREW CORPORATION	10500 W 153rd Street Orland Park, Illinois 60462
CALIFORNIA MICROWAVE INC.	990 Almanor Avenue Sunnyvale, California 94086
COMSAT TECHNOLOGY PRODUCTS	22300 Comsat Drive Clarksburg, Maryland 20871
COMTECH ANTENNA CORP.	3100 Communications Road St Cloud, Florida 32769
GTE INTERNATIONAL SYSTEMS CORP.	140 First Avenue Waltham, Massachussetts 02254
HARRIS SATELLITE COMMUNICATION DIV.	PO Box 1700 Melbourne, Florida 32901
M/A-COM TELECOMMUNICATION DIV.	3033 Science Park Road San Diego, California 92121
MICRODYNE CORP.	PO Box 7213 Ocala, Florida 32672

SCIENTIFIC-ATLANTA INC.	3845 Pleasantdale Road Atlanta, Georgia 30340
VERTEX COMMUNICATIONS CORP.	2600 Longview Street Kilgore, Texas 75662

Owners of international communications satellites

INTELSAT (World)	3400 International Drive NW Washington DC 20546 USA
INMARSAT (Maritime)	40 Melton Street London NW1 2EQ England
EUTELSAT (Europe)	Tour Maine-Montparnasse 33 avenue du Maine 75755 Paris Cedex 15 France

Owners of domestic communications satellites – USA

AMERICAN SATELLITE CO.	1801 Research Boulevard Rockville, Maryland 20850
AT&T COMMUNICATIONS	295 N. Maple Avenue Basking Ridge, New Jersey 07920
FORD AEROSPACE	1140 Connecticut Avenue NW Washington DC 20036
GTE SPACENET CORP.	1700 Old Meadow Road McLean, Virginia 22102
HUGHES COMMUNICATIONS	PO Box 92424 Los Angeles, California 90009
RCA AMERICOM	400 College Road E. Princeton, New Jersey 08540
SATELLITE BUSINESS SYSTEMS	8283 Greensboro Drive McLean, Virginia 22102
WESTERN UNION	One Lake Street Upper Saddle River, New Jersey 07920

Owners of other domestic and regional communications satellites

AUSTRALIA	Aussat Pty Ltd 54 Carrington Street Sydney 2000
BRAZIL	Embratel Avenida Presidente Vargas 1012 20–071 Rio de Janeiro
CANADA	Telesat 333 River Road Ottawa, Ontario K1L 8B9
FRANCE	Direction Générale des Télécommunications 38–40 rue du Général-Leclerc 92131 Issy Les Moulineaux
INDONESIA	Perumtel Jalan Cisanggarung 2 Bandung
JAPAN	Ministry of Posts and Telecommunications 3–2 Kasumigaseki 1-Chome Chiyoda-ku Tokyo 100
LUXEMBOURG (Planned)	Societé Européenne des Satellites (SES) 63 Avenue de la Liberté L-1931 Luxembourg
MEXICO	Direccion General de Telecommunicaciones Ave. Lazaro Cardenas 567 03028 Mexico DF
SAUDI ARABIA	Arabsat PO Box 1038 Riyadh
UK (Planned)	BSB Vision House 19–72 Rathbone Place London W1P 1DF

Index